Dynamical Systems

Werner Krabs • Stefan Pickl

Dynamical Systems

Stability, Controllability and Chaotic Behavior

Springer

Prof. Dr. Werner Krabs
Department of Mathematics
Technical University of
Darmstadt
Schlossgartenstr. 7
64289 Darmstadt
Germany

Prof. Dr. Stefan Pickl
Universität der Bundeswehr
München
Department of Computer Science
Werner Heisenberg Weg 39
85577 Neubiberg-München
Germany
stefan.pickl@unibw.de

ISBN 978-3-642-13721-1 e-ISBN 978-3-642-13722-8
DOI 10.1007/978-3-642-13722-8
Springer Heidelberg Dordrecht London New York

Library of Congress Control Number: 2010932600

Cover design: WMXDesign GmbH, Heidelberg, Germany

Printed on acid-free paper

Springer is part of Springer Science+Business Media (www.springer.com)

Foreword

Reflecting modelling dynamical systems by mathematical methods can be enriched by philosophical categories. The following introduction might catch the reader's interest concerning some interdisciplinary dimensions and completes the holistic approach.

"It has been said– and I was among those saying it– that any theory of explanation worth its salt should be able to make good predictions. If good predictions could not be made, the explanation could hardly count as serious. (...) I now want to move on to my main point. There is, I claim, no major conceptual difference between the problems of explaining the unpredictable in human affairs and in non-human affairs. There are, it is true, many remarkable successes of prediction in the physical sciences of which we are all aware, but these few successes of principled science making principled predictions are, in many ways, misleading. (...) The point I want to emphasize is that instability is as present in purely physical systems as it is in those we think of as characteristically human. Our ability to explain but not predict human behavior is in the same general category as our ability to explain but not predict many physical phenomena. The underlying reasons for the inability to predict are the same. (...) The concept of instability which accounts for many of these failures is one of the most neglected concepts in philosophy. We philosophers have as a matter of practice put too much emphasis on the contrast between deterministic and probabilistic phenomena. We have not emphasized enough the salient differences between stable and unstable phenomena. One can argue that the main sources of probabilistic or random behavior lie in instability. We might even want to hold the speculative thesis that the random behavior of quantum systems will itself in turn be explained by unstable behavior of classical dynamical systems. (...)

Chaos, the original confusion in which all the elements were mixed together, was personified by the Greeks as the most ancient of the gods. Now in the twentieth century, chaos has returned in force to attack that citadel of order and harmony, classical mechanics. We have come to recognize how rare and special are those physical systems whose behavior can be predicted in detail. The naivet and hopes of earlier years will not return. For many phenomena in many domains there are principled reasons to believe that we shall never be able to move from good explanations to good predictions."

PATRICK SUPPES "EXPLAINING THE UNPREDICTABLE"
Lucie Stern Professor of Philosophy, Stanford University
Director and Faculty Advisor, Education Program for Gifted Youth, Stanford University

Preface

At the end of the nineteenth century Lyapunov and Poincaré developed the so called qualitative theory of differential equations and introduced geometric-topological considerations which have led to the concept of dynamical systems. In its present abstract form this concept goes back to G.D. Birkhoff.

This is also the starting point of Chapter 1 of this book in which uncontrolled and controlled time-continuous and time-discrete systems are investigated under the aspect of stability and controllability. Chapter 1 starts with time-continuous dynamical systems. After the description of elementary properties of such systems it focusses on stability in the sense of Lyapunov and gives applications to systems in the plane such as the mathematical pendulum, to general predator-prey models, and to evolution matrix games.

The time-discrete case is divided into the autonomous and the non-autonomous part where the latter is no more a dynamical system in the strong sense. It is the counter part of the time-continuous case where the right-hand side of the system of differential equations which describes the dynamics of the system depends explicitly on the time.

Controlled dynamical systems could be considered as dynamical systems in the strong sense, if the controls were incorporated into the state space. We, however, adopt the conventional treatment of controlled systems as in control theory. We are mainly interested in the question of controllability of dynamical systems into equilibrium states. In the non-autonomous time-discrete case we also consider the problem of stabilization.

Chapter 3 is concerned with chaotic behaviour of autonomous time discrete systems. We consider three different types of chaos: chaos in the sense of Devaney, disorder chaos and chaos in the sense of Li and Yorke. The chapter ends with two examples of strange (or chaotic) attractors.

The Appendix A is concerned with a dynamical method for the calculation of Nash equilibria in non-cooperative n-person games. The method is based on the fact that Nash equilibria are fixed points of certain continuous mappings of the Cartesian product of the strategy sets of the players into itself. This gives rise to an iteration method for the calculation of Nash equilibria the set of which can be considered as the Ω-limit set of a time-discrete dynamical system.

In Appendix B we consider two optimal control problems in chemotherapeutic treatment of cancer. These two problems are somehow dual to each other and are shown to have solutions of the same type.

The authors want to thank Korcan Görgülü for excellent typesetting. He solved every TEX- problem which occurred in minimal time.

Werner Krabs, Stefan Pickl, Munich, May 2010

Contents

1 Uncontrolled Systems ... 1
1.1 Abstract Definition of Dynamical Systems ... 1
1.2 Time-Continuous Dynamical Systems ... 4
1.2.1 Elementary Properties ... 4
1.2.2 Systems in the Plane ... 8
1.2.3 Stability: The Direct Method of Lyapunov ... 14
1.2.4 Application to a General Predator-Prey Model ... 21
1.2.5 Application to Evolution Matrix Games ... 32
1.3 Time-Discrete Dynamical Systems ... 36
1.3.1 The Autonomous Case: Definitions and Elementary Properties ... 36
1.3.2 Localization of Limit Sets with the Aid of Lyapunov Functions ... 40
1.3.3 Stability Based on Lyapunov's Method ... 42
1.3.4 Stability of Fixed Points via Linearisation ... 47
1.3.5 Linear Systems ... 50
1.3.6 Discretization of Time-Continuous Dynamical Systems . 55
1.3.7 Applications ... 57
1.3.8 The Non-Autonomous Case: Definitions and Elementary Properties ... 62
1.3.9 Stability Based on Lyapunov's Method ... 64
1.3.10 Linear Systems ... 67
1.3.11 Application to a Model for the Process of Hemo-Dialysis 73

2 Controlled Systems ... 77
2.1 The Time-Continuous Case ... 77
2.1.1 The Problem of Controllability ... 77
2.1.2 Controllability of Linear Systems ... 78
2.1.3 Restricted Null-Controllability of Linear Systems ... 82
2.1.4 Controllability of Nonlinear Systems into Rest Points .. 85

2.1.5 Approximate Solution of the Problem of Restricted Null-Controllability 89
2.1.6 Time-Minimal Restricted Null-Controllability of Linear Systems 91
2.2 The Time-Discrete Autonomous Case 101
2.2.1 The Problem of Fixed Point Controllability 101
2.2.2 Null-Controllability of Linear Systems 111
2.2.3 A Method for Solving the Problem of Null-Controllability 118
2.2.4 Stabilization of Controlled Systems 124
2.2.5 Applications 126
2.3 The Time-Discrete Non-Autonomous Case 134
2.3.1 The Problem of Fixed Point Controllability 134
2.3.2 The General Problem of Controllability 138
2.3.3 Stabilization of Controlled Systems 141
2.3.4 The Problem of Reachability 143

3 Chaotic Behavior of Autonomous Time-Discrete Systems .. 149
3.1 Chaos in the Sense of Devaney 149
3.2 Topological Conjugacy 154
3.3 The Topological Entropy as a Measure for Chaos 163
3.3.1 Definition and Invariance 163
3.3.2 The Topological Entropy of the Shift-Mapping 165
3.3.3 Disorder-Chaos for One-Dimensional Mappings 167
3.4 Chaos in the Sense of Li and Yorke 172
3.5 Strange (or Chaotic) Attractors 180
3.6 Bibliographical Remarks 190

A A Dynamical Method for the Calculation of Nash-Equilibria in $n-$Person Games 195
A.1 The Fixed Point Theorems of Brouwer and Kakutani 195
A.2 Nash Equilibria as Fixed Points of Mappings 197
A.3 Bi-Matrix Games 202
A.4 Evolution Matrix Games 209

B Optimal Control in Chemotherapy of Cancer 217
B.1 The Mathematical Model and Two Control Problems 217
B.2 Solution of the First Control Problem 220
B.3 Solution of the Second Control Problem 224
B.4 Pontryagin's Maximum Principle 229

C List of Authors 231

References 233

Index 235

1

Uncontrolled Systems

1.1 Abstract Definition of Dynamical Systems

The concept of dynamical systems has developed out of the qualitative theory of differential equations which was established by Lyapunov and Poincaré in the course of the two last decades of the nineteenth century. As final result of a development which lasted more than half a century the following abstract definition of a dynamical system has grown out:

Let X be a metric space with metric d. Further let I be an additive semigroup of real numbers, i.e. a subset I of $\mathbb{R}$ with

$$\begin{aligned} 0 &\in I, \\ t, s \in I \quad &\Longrightarrow \quad t + s = s + t \in I, \\ t, s, r \in I \quad &\Longrightarrow \quad (t + s) + r = t + (s + r). \end{aligned}$$

A dynamical system on X, sometimes also called a flow, is then defined by a continuous mapping $\pi : X \times I \to X$ with the following properties:

(a) $\pi(x, 0) = x$ for all $x \in X$ (identity property),

(b) $\pi(\pi(x, t), s) = \pi(x, t + s)$ for all $x \in X$ (semigroup property).

The metric space X is the space of states of the system and the mapping π describes the temporal change of the system where the time t proceeds in a semigroup $I \subseteq \mathbb{R}$.

As a rule we have $I = \mathbb{R}$, $I = \mathbb{R}_+ = \{t \in \mathbb{R} \mid t \geq 0\}$ or $I = \mathbb{N}_0 = \{0, 1, 2, \ldots\}$. In the latter case the dynamical system is called time-discrete. Historically the first example of a dynamical system is the following:

S. Pickl and W. Krabs, *Dynamical Systems: Stability, Controllability and Chaotic Behavior*,
DOI 10.1007/978-3-642-13722-8_1,

Let W be a non-empty, open and connected subset of $\mathbb{R}^r$ and let $f : W \to \mathbb{R}^r$ be a Lipschitz-continuous mapping, i.e., there exists a constant $K > 0$ such that

$$||f(x) - f(y)||_2 \leq K||x - y||_2 \qquad \text{for all} \quad x, y \in W$$

where $||\cdot||_2$ denotes the Euclidian norm in $\mathbb{R}^r$. Then, for every $x_0 \in W$, the initial value problem

$$\dot{x}(t) = f(x(t)), \quad x(0) = x_0$$

has a unique solution $x(t) = \varphi(x_0, t)$ which is defined on an open interval $-\alpha < t < \alpha$ for some $\alpha > 0$ and which belongs to $\mathcal{C}^1((-\alpha, \alpha), W)$. We assume $\alpha = \infty$ and put $I = \mathbb{R}$, $X = W$. We equip X with the metric

$$d(x, y) = ||x - y||_2, \qquad x, y \in I.$$

Then we define $\pi : X \times I \to X$ by

$$\pi(x, t) = \varphi(x, t) \qquad \text{for all} \quad x \in W \quad \text{and} \quad t \in I.$$

Obviously we have

$$\pi(x, 0) = \varphi(x, 0) = x \qquad \text{for all} \quad x \in X,$$

i.e., π has the identity property.

If $t \in I$ is given, then we define, for every $x \in X$:

$$\widetilde{\varphi}(x, s) = \varphi(x, t + s) \qquad \text{for all} \quad s \in I$$

and conclude

$$\dot{\widetilde{\varphi}}(x, s) = \dot{\varphi}(x, t + s) = f(x(t + s)) = f\left(\widetilde{\varphi}(x, s)\right)$$

and

$$\widetilde{\varphi}(x, 0) = \varphi(x, t).$$

This implies

$$\pi(\pi(x, t), s) = \pi(\varphi(x, t), s) = \widetilde{\varphi}(x, s) = \varphi(x, t + s) = \pi(x, t + s),$$

i.e., π also has the semigroup property.

The continuity of the mapping $\pi : X \times I \to X$ is a consequence of a well known theorem about the continuous dependence of the solutions of initial value problems on the initial values.

Let us return to the general definition of a dynamical system and define an orbit or a trajectory through $x \in X$ by the point set

$$\gamma_I(x) = \{\pi(x, t) \mid t \in I\} = \bigcup_{t \in I} \{\pi(x, t)\}. \qquad (1.1)$$

Obviously we have

$$x = \pi(x, 0) \in \gamma_I(x).$$

For every $x \in X$ we then define the limit set by

$$L_I(x) = \bigcap_{t \in I} \overline{\gamma_I(\pi(x,t))} \tag{1.2}$$

where for an arbitrary subset A of X the set $\overline{A}$ denotes the closure of A.

1.2 Time-Continuous Dynamical Systems

1.2.1 Elementary Properties

Let $I = \mathbb{R}$. Then the dynamical system is called time-continuous. For every $x \in X$ the positive and negative halftrajectory is defined by

$$\gamma_+(x) = \{\pi(x,t) \mid t \geq 0\} \qquad \text{and} \qquad \gamma_-(x) = \{\pi(x,t) \mid t \leq 0\}, \tag{1.3}$$

respectively. Further the omega and alpha limit set is defined by

$$\Omega(x) = \bigcap_{t\geq 0} \overline{\gamma_+(\pi(x,t))} \qquad \text{and} \qquad A(x) = \bigcap_{t\leq 0} \overline{\gamma_-(\pi(x,t))}. \tag{1.4}$$

The omega limit set is of primary interest, because it describes the asymptotic behavior of the dynamical system for $t \to \infty$. In this case we can prove the

Lemma 1.1. *For every $x \in X$ we have*

$$\Omega(x) = \Big\{ y \in X \mid y = \lim_{n\to\infty} \pi(x,t_n) \text{ for a sequence } (t_n)_{n\in\mathbb{N}_0} \text{ in } \mathbb{R}_+ \text{ with } t_n \to \infty \Big\}. \tag{1.5}$$

Proof.
1) Let $y \in X$ be such that $y = \lim_{n\to\infty} \pi(x,t_n)$ for a sequence $(t_n)_{n\in\mathbb{N}_0}$ in $\mathbb{R}_+$ with $t_n \to \infty$. Then it follows that $y \in \overline{\gamma_+(x)}$. However, for every $t \geq 0$ we also have

$$y = \lim_{n\to\infty} \pi(x,t_n) = \lim_{n\to\infty} \pi(\pi(x,t), t_n - t),$$

since $t_n - t \geq 0$ for sufficiently large $n \in \mathbb{N}_0$ and $t_n - t \to \infty$. Hence it follows that $y \in \overline{\gamma_+(\pi(x,t))}$ for every $t \geq 0$ which implies $y \in \Omega(x)$ defined by (1.4).
2) Conversely let $y \in \Omega(x)$ be defined by (1.4). Then

$$y \in \overline{\gamma_+(\pi(x,t))} \qquad \text{for every} \quad t \geq 0$$

and there exists a sequence $(x_n)_{n\in\mathbb{N}_0}$ in $\gamma_+(\pi(x,t))$ with $y = \lim_{n\to\infty} x_n$.

For every $n \in \mathbb{N}_0$ we can also write

$$x_n = \pi(x, t + t_n) \qquad \text{for some} \quad t \geq 0$$

such that

$$y = \lim_{n\to\infty} \pi(x, t + t_n) \qquad \text{for all} \quad t \geq 0.$$

For every $k \in \mathbb{N}$ therefore there exists some $n_k \in \mathbb{N}$ with

$$d(y, \pi(x, k + t_n)) \leq \frac{1}{k} \qquad \text{for all} \quad n \geq n_k.$$

This implies

$$y = \lim_{k\to\infty} \pi(x, k + t_{n_k}),$$

i.e., y belongs to $\Omega(x)$ defined by (1.5). □

Definition. *A subset $A \subseteq X$ is called invariant, if*

$$\gamma_{\mathbb{R}}(x) \subseteq A \qquad \text{for all} \quad x \in A$$

where $\gamma_{\mathbb{R}}(x)$, for every $x \in A$, is the trajectory defined by (1.1).

In words: $A \subseteq X$ is called invariant, if every trajectory through a point $x \in A$ is contained in A.

Lemma 1.2. *$A \subseteq X$ is invariant, if and only if*

$$A = \bigcup_{x \in A} \gamma_{\mathbb{R}}(x). \tag{1.6}$$

Proof.
1) Let A be invariant. Then

$$\bigcup_{x \in A} \gamma_{\mathbb{R}}(x) \subseteq A \subseteq \bigcup_{x \in A} \gamma_{\mathbb{R}}(x), \text{ i.e. (1.6) holds true.}$$

2) Conversely it follows from (1.6) that

$$\gamma_{\mathbb{R}}(x) \subseteq A \qquad \text{for all} \quad x \in A, \text{ i.e., } A \text{ is invariant.}$$ □

Proposition 1.3. *The closure of an invariant subset of X is also invariant.*

Proof. Let $A \subseteq X$ be invariant and let $y \in \overline{A}$. Then we have to show that $\gamma_{\mathbb{R}}(y) \subseteq \overline{A}$. Since $y \in \overline{A}$ there is a sequence $(y_n)_{n \in \mathbb{N}}$ in A with $y = \lim_{n \to \infty} y_n$. For every $t \in \mathbb{R}$ it therefore follows because of the continuity of the mapping $\pi : X \times \mathbb{R} \to X$ that

$$\pi(y, t) = \lim_{n \to \infty} \pi(y_n, t).$$

Because of $\gamma_{\mathbb{R}}(y_n) \subseteq A$ we have $\pi(y_n, t) \in A$ for all n and therefore $\pi(y, t) \in \overline{A}$ for every $t \in \mathbb{R}$ which implies that $\gamma_{\mathbb{R}}(x) \subseteq \overline{A}$. □

Proposition 1.4. *The omega limit set $\Omega(x)$ is closed and invariant for every $x \in X$.*

Proof. For every $x \in X$ the set $\Omega(x)$ is closed as the intersection of closed sets. In order to prove invariance of $\Omega(x)$ we choose some $y \in \Omega(x)$. Then because of Lemma 1.1 there is a sequence $(t_n)_{n \in \mathbb{N}_0}$ in $\mathbb{R}_+$ with $t_n \to \infty$ such that $y = \lim_{n \to \infty} \pi(x, t_n)$. If we put

$$x_n = \pi(x, t_n) \qquad \text{for every} \quad n \in \mathbb{N}_0,$$

then it follows $y = \lim_{n \to \infty} x_n$ and the continuity of $\pi : X \times \mathbb{R} \to X$ implies

$$\pi(y, t) = \lim_{n \to \infty} \pi(x_n, t) \qquad \text{for all} \quad t \in \mathbb{R}.$$

Now we have

$$\pi(x_n, t) = \pi(\pi(x, t_n), t) = \pi(x, t_n + t) \quad \text{and} \quad t_n + t \to \infty$$

for every $t \in \mathbb{R}$. Therefore it follows that

$$\pi(y,t) \in \Omega(x) \qquad \text{for all} \quad t \in \mathbb{R} \quad \text{and hence} \quad \gamma_{\mathbb{R}}(y) \subseteq \Omega(x)$$

which shows that $\Omega(x)$ is invariant. □

For every $x \in X$ we call the function $t \mapsto \pi(x,t)$, $t \in \mathbb{R}$, a movement through x.

Definition. *A movement $\pi(x,t)$, $t \in \mathbb{R}$, through $x \in X$ is called positively compact, if the closure of the positive halftrajectory $\gamma_+(x)$ defined by (1.3) is compact which is equivalent to the existence of a compact subset K of X with $\pi(x,t) \in K$ for all $t \geq 0$.*

With this definition we can prove the following

Proposition 1.5. *Let $\pi(x,t)$, $t \in \mathbb{R}$, for some $x \in X$, be a positively compact movement. Then the omega limit set $\Omega(x)$ defined by (1.4) is non-empty, compact, invariant and connected.*

Proof. By Proposition 1.4 the set $\Omega(x)$ is invariant. By assumption $\overline{\gamma_+(x)}$ with $\gamma_+(x) = \{\pi(x,t) \mid t \geq 0\}$ is compact. Further we have

$$\overline{\gamma_+(\pi(x,t))} \subseteq \overline{\gamma_+(x)} \qquad \text{for all} \quad t \geq 0.$$

Hence

$$\overline{\gamma_+(\pi(x,t))} \qquad \text{is compact for all} \quad t \geq 0.$$

Finally we have

$$\overline{\gamma_+(\pi(x,t_2))} \subseteq \gamma_+(\pi(x,t_1)) \qquad \text{for} \quad t_2 \geq t_1 \geq 0$$

which implies that $\Omega(x)$ is non-empty and compact.

Let us assume that $\Omega(x)$ is not connected. Then there exist two disjoint, non-empty, open subsets A and B of X with $\Omega(x) = \Omega(x) \cap (A \cup B)$ where $A \cap \Omega(x)$ and $B \cap \Omega(x)$ are non-empty and we can find sequences $(s_n)_{n \in \mathbb{N}}$ and $(t_n)_{n \in \mathbb{N}}$ in $\mathbb{R}_+$ such that

$$0 < s_1 < t_1 < \ldots < s_n < t_n < \ldots, \; s_n \to \infty, \; t_n \to \infty,$$
$$\pi(x, s_n) \in A \quad \text{and} \quad \pi(x, t_n) \in B \quad \text{for all} \quad n \in \mathbb{N}.$$

Since, for every $n \in \mathbb{N}$, the path $\pi(x, [s_n, t_n])$ is connected, there is, for every $n \in \mathbb{N}$, some $t_n^* \in [s_n, t_n]$ with $\pi(x, t_n^*) \in A \cup B$. From $\pi(x, [s_n, t_n]) \subseteq \gamma_+(x)$ and the compactness of $\overline{\gamma_+(x)}$ it follows that there exists a subsequence $\left(\pi\left(x, t_{n_i}^*\right)\right)_{i \in \mathbb{N}}$ and a point $y \in X \setminus (A \cup B)$ with $y = \lim_{i \to \infty} \pi(x, t_{n_i})$, i.e. $y \in \Omega(x)$ by Lemma 1.1. However, this is impossible because of $\Omega(x) = \Omega(x) \cap (A \cup B)$. This contradiction shows that $\Omega(x)$ is connected. □

Definition. *A subset $M \subseteq X$ is called minimal, if it is non-empty, closed and invariant and none of its genuine subsets has these properties.*

Proposition 1.6. *Every non-empty, compact and invariant subset A of X contains a minimal subset M of X.*

Proof. Let $\mathcal{A}$ be the set of all non-empty, compact and invariant subsets of A. The set A is semi-ordered with respect to inclusion. Now let $\mathcal{B}$ a well ordered subset of $\mathcal{A}$. Then the intersection $\bigcap\limits_{B\in\mathcal{B}} B$ is non-empty, in $\mathcal{A}$ and a lower bound of $\mathcal{B}$. By Zorn's Lemma it follows that $\mathcal{A}$ possesses minimal elements.
□

As an immediate consequence of Proposition 1.5 and Proposition 1.6 we obtain the

Corollary. *If $\pi(x,t)$, $t \in \mathbb{R}$, for some $x \in X$, is a positively compact movement through x, then the omega limit set $\Omega(x)$ contains a minimal subset.*

Definition. *A point $x \in X$ is called a rest point or equilibrium of a flow $\pi : X \times \mathbb{R} \to X$, if*

$$\pi(x,t) = x \qquad \text{for all} \quad t \in \mathbb{R}.$$

If $x \in X$ is a rest point of a flow, then $M = \{x\}$ is obviously minimal.

Definition. *A point $x \in X$ and the movement $\pi(x,t)$, $t \in \mathbb{R}$, through x is called periodic, if there exists some $p > 0$ with*

$$\pi(x,t+p) = \pi(x,t) \qquad \textit{for all} \quad t \in \mathbb{R}.$$

The number p is called the period of the movement $\pi(x,t)$, $t \in \mathbb{R}$, through x.

Proposition 1.7. *If $x \in X$ is periodic, then the trajectory $\gamma_{\mathbb{R}}(x)$ is minimal.*

Proof. The periodicity of $x \in X$ implies $\gamma_{\mathbb{R}}(x) = \bigcup\limits_{t\in[0,p]} \pi(x,t)$. Therefore, $\gamma_{\mathbb{R}}(x)$ is compact, hence closed.

Now let $y \in \gamma_{\mathbb{R}}(x)$. Then there is some $t_y \in [0,p]$ with $y = \pi(x,t_y)$ and for every $s \in \mathbb{R}$ it follows that

$$\pi(y,s) = \pi\left(\pi\left(x,t_y\right),s\right) = \pi\left(x,s+t_y\right) \in \gamma_{\mathbb{R}}(x),$$

hence, $\gamma_{\mathbb{R}}(y) \subseteq \gamma_{\mathbb{R}}(x)$ which proves the invariance of $\gamma_{\mathbb{R}}(x)$.

Now let $z \in \gamma_{\mathbb{R}}(x)$ be chosen arbitrarily. Then there is some $t_z \in [0,p]$ with $z = \pi(x,t_z)$. This implies that

$$z = \pi\left(x,t_z+t_y-t_y\right) = \pi\left(\pi\left(x,t_y\right),t_z-t_y\right) = \pi\left(y,t_z-t_y\right) \in \gamma_{\mathbb{R}}(y),$$

hence, $\gamma_{\mathbb{R}}(x) \subseteq \gamma_{\mathbb{R}}(y)$, which implies

$$\gamma_{\mathbb{R}}(x) = \gamma_{\mathbb{R}}(y) \qquad \text{for all} \quad y \in \gamma_{\mathbb{R}}(x).$$

Therefore, there cannot exist a genuine non-empty, closed subset of $\gamma_{\mathbb{R}}(x)$ which is invariant. This shows the minimality of $\gamma_{\mathbb{R}}(x)$. □

1.2.2 Systems in the Plane

We start with the historically first example of a dynamical system for the case $r = 2$ (see Section 1.1). This is defined by two differential equations of the form

$$\left.\begin{aligned}\dot{x}_1 &= f_1(x_1, x_2),\\ \dot{x}_2 &= f_2(x_1, x_2),\end{aligned}\right\} \qquad (x_1, x_2) \in W \subseteq \mathbb{R}^2, \tag{1.7}$$

where W is non-empty, open and connected. We assume $f_1, f_2 \in \mathcal{C}^1(W)$ which implies that for every point $(x_{10}, x_{20}) \in W$ there is exactly one solution $x(t) = \varphi(t, x_0)$, $t \in (-\alpha, \alpha)$ for some $\alpha > 0$, of (1.7) with $\varphi(\cdot, x_0) \in \mathcal{C}^1((-\alpha, \alpha), W)$ and

$$\varphi(0, x_0) = x_0 = (x_{10}, x_{20}). \tag{1.8}$$

If $X = W$ is equipped with the metric

$$d(x, y) = ||x - y||_2, \qquad x, y \in W,$$

where $||\cdot||_2$ denotes the Euclidian norm in $\mathbb{R}^2$ and if we assume $\alpha = \infty$, then (as been shown in Section 1.1) by

$$\pi(x, t) = \varphi(t, x), \qquad t \in \mathbb{R},\ x \in X, \tag{1.9}$$

a (time-continuous) dynamical system $\pi : X \times \mathbb{R} \to X$ is defined.

A point $x \in X$ is a rest point of the flow π, if

$$f_1(x) = 0 \qquad \text{and} \qquad f_2(x) = 0.$$

We already observed that every rest point of π as a one point subset of X is minimal. The question arises then whether there exist further minimal sets. For the case that these are compact the following proposition gives a complete answer.

Proposition 1.8. *Let $M \subseteq X$ be compact and minimal. Then either M is a rest point of the flow (1.9) or a periodic movement $\pi(x, t)$, $t \in \mathbb{R}_+$, through some $x \in M$.*

In the proof of this proposition we need the following

Proposition 1.9. *If, for some $x \in X$, the positive orbit $\gamma_+(x)$ of (1.3) and the omega limit set $\Omega(x)$ have non-rest points in common, then $\gamma_+(x)$ is a periodic orbit, i.e., there exists some $p > 0$ such that*

$$\pi(x, t + p) = \pi(x, t) \qquad \textit{for all} \quad t \geq 0.$$

For the proof of this proposition we refer to the book "Ordinary Differential Equations" by J.K. Hale (Wiley-Interscience 1969).

Proof. Choose $x \in M$ arbitrarily. Then $\overline{\gamma_+(x)} \in M$. By Proposition 1.4 and Proposition 1.5 the omega limit set $\Omega(x)$ is in M, non-empty, closed and invariant which implies, by the minimality of M, that $M = \Omega(x)$. If M contains a rest point y, then it follows that $M = \{y\} = \{x\}$. If M does not contain a rest point, then it follows from $\gamma_+(x) \subseteq M = \Omega(x)$ that $\gamma_+(x)$ and $\Omega(x)$ have common non-rest points so that according to Proposition 1.9 $\gamma_+(x)$ is a periodic orbit. □

Further we have

Proposition 1.10. *Let $\gamma_{\mathbb{R}}(x)$, for some $x \in X$, be a positively compact orbit, i.e., the movement $\pi(x,t)$, $t \in \mathbb{R}$, through x is positively compact. Further, let the omega limit set $\Omega(x)$ contain no rest point. Then $\Omega(x) = \gamma_+(x)$ is a periodic orbit.*

For the proof of this proposition we need the following

Proposition 1.11. *Does $\Omega(x)$, for some $x \in X$, contain non-rest points and a periodic orbit $\gamma_+(x)$ then $\Omega(x) = \gamma_+(x)$.*

For the proof of this proposition we again refer to the book by J.K. Hale mentioned above.

Proof. By Proposition 1.5 $\Omega(x)$ is non-empty, compact and invariant. Therefore, $\Omega(x)$ contains a minimal subset M of X and M does not contain rest points. By Proposition 1.8 M is a periodic movement $\pi(x,t)$, $t \in \mathbb{R}_+$, through x. By Proposition 1.11, therefore, $\Omega(x) = \gamma_+(x)$ is a periodic orbit. □

Let us demonstrate Proposition 1.10 by two examples

1) The Mathematical Pendulum

We consider the movement of a nonlinear plane pendulum with a fixed suspension point as being depicted in the following picture:

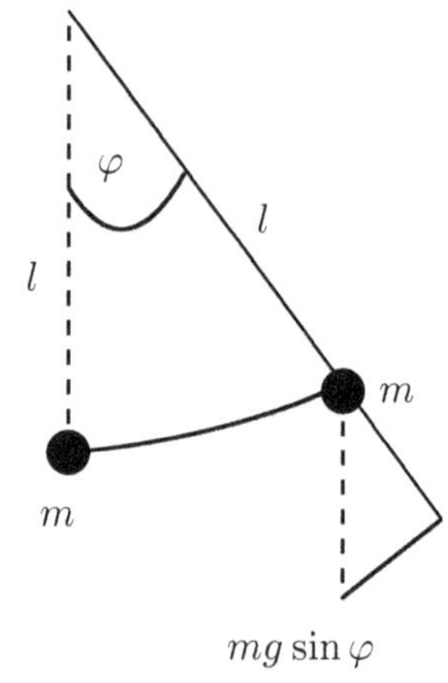

By Newton's law

$$force = mass \times acceleration$$

we obtain the following equation of movement

$$m \cdot l \cdot \ddot{\varphi}(t) = -mg \sin \varphi(t), \qquad t \in \mathbb{R}, \tag{1.10}$$

where m and l is the mass and the length of the pendulum, respectively, g is the gravity constant and $\varphi(t)$ the vibration angle of the time t. The expression $l \cdot \ddot{\varphi}(t)$ is the acceleration along the orbit and $-mg \sin \varphi(t)$ the force upon the mass of the pendulum at the time t.

If we define

$$x_1(t) = \varphi(t) \quad \text{and} \quad x_2(t) = \dot{\varphi}(t), \qquad t \in \mathbb{R}, \tag{1.11}$$

then (1.10) turns out to be equivalent to

$$\begin{aligned} \dot{x}_1(t) &= x_2(t), \\ \dot{x}_2(t) &= -\frac{g}{l} \sin x_1(t), \qquad t \in \mathbb{R}. \end{aligned} \tag{1.12}$$

With $f_1(x_1, x_2) = x_2, \; f_2(x_1, x_2) = -\frac{g}{l} \cdot \sin x_1, \; (x_1, x_2) \in \mathbb{R}^2$, the system (1.12) takes the form of (1.7) with $W = \mathbb{R}^2$. Obviously $f_1, f_2 \in C^1(W)$ such that for every point $(x_{10}, x_{20}) = (\varphi_0, \dot{\varphi}_0) \in \mathbb{R}^2$ there exists exactly one solution $(x_1(\cdot), x_2(\cdot)) \in C^1((-\alpha, \alpha), W)$ of (1.12) for some $\alpha > 0$ with

$$x_1(0) = x_{10}, \qquad x_2(0) = x_{20}. \tag{1.13}$$

With (1.11) it follows from (1.10) that

$$\ddot{x}_1 = -\frac{g}{l}\sin x_1$$

which implies that

$$\dot{x}_1^2 = \frac{2g}{l}\cos x_1 + C \qquad \text{for some} \quad C \in \mathbb{R}$$

such that with (1.13) we obtain

$$x_{20}^2 = \frac{2g}{l}\cos x_{10} + C.$$

This leads to

$$x_2^2 = \dot{x}_1^2 = x_{20}^2 + \frac{2g}{l}(\cos x_1 - \cos x_{10}) \leq x_{20}^2 + \frac{4g}{l}$$

and finally to

$$|x_2| \leq \sqrt{x_{20}^2 + \frac{4g}{l}}.$$

If we choose $X = (-\pi, \pi) \times \mathbb{R}$, then for every point $(x_{10}, x_{20}) \in X$ the unique solution $(x_1(\cdot), x_2(\cdot)) \in C^1((-\alpha, \alpha), X)$ of (1.12) with (1.13) satisfies

$$|x_1(\cdot)| \leq \pi, \qquad |x_2| \leq \sqrt{x_{20}^2 + \frac{4g}{l}}$$

so that we can choose $\alpha = \infty$.

If one defines, for every $(x_{10}, x_{20}) \in X$ and every $t \in \mathbb{R}$

$$\pi(x_{10}, x_{20}, t) = (x_1(t), x_2(t)), \qquad t \in \mathbb{R},$$

where $(x_1(\cdot), x_2(\cdot))$ is the unique solution of (1.12), (1.13) in $C^1(\mathbb{R}, X)$, then one obtains a flow such that for every $x \in X$ the movement $\pi(x,t)$, $t \in \mathbb{R}$, through x is positively compact.

Obviously $(0,0)$ is the only rest point of (1.12) in X. Now let $y \in \Omega(x)$ be given for some $x \in X$. By Lemma 1.1 there exists a sequence $(t_n)_{n\in\mathbb{N}_0}$ in $\mathbb{R}_+$ with $t_n \to \infty$ and $y = \lim_{n\to\infty} \pi(x, t_n)$. This implies

$$y_2^2 = x_2^2 + \frac{2g}{l}(\cos y_1 - \cos x_1).$$

If y were $(0,0)$, then it would follow that

$$0 = x_2^2 + \frac{2g}{l}(1 - \cos x_1)$$

which is only possible in the case $x_1 = x_2 = 0$. Therefore, if $x = (0,0)$, no $y \in \Omega(x)$ can be $y = (0,0)$, i.e., $\Omega(x)$ does not contain rest points. By Proposition 1.10 it therefore follows that $\Omega(x) = \gamma_+(x)$ is a periodic orbit.

2) A Predator-Prey Model

This model is described by the differential equations

$$\begin{aligned} \dot{x}_1 &= -c\, x_1 + a\, x_1 x_2, \\ \dot{x}_2 &= r\, x_2 - p\, x_1 x_2, \qquad x_1 > 0,\ x_2 > 0, \end{aligned} \tag{1.14}$$

where $a, c, p, r \in \mathbb{R}$ are given positive constants. By $x_1(t)$ and $x_2(t)$ the size of a predator and prey population, respectively, at time t within a given habitat are denoted. The first equation of (1.14) describes the temporal change of the size of the predator and the second equation that of the prey population. In this case we have

$$X = W = \{(x_1, x_2) \in \mathbb{R}^2 \mid x_1 > 0 \text{ and } x_2 > 0\}$$

and all the assumptions we have made with respect to (1.7) are satisfied.

We shall show that every solution of (1.14) is bounded and hence defined on all of $\mathbb{R}$ such that (1.14) defines a flow $\pi : X \times \mathbb{R} \to X$. The only rest point of this flow is given by

$$\overline{x}_1 = \frac{r}{p}, \qquad \overline{x}_2 = \frac{c}{a}.$$

In order to apply Proposition 1.10 we at first show that for every $x = (x_1, x_2) \in X$ the orbit $\gamma_{\mathbb{R}}(x)$ is positively compact. For that purpose we consider the homeomorphic mapping $T : X \to \mathbb{R}^2$ which is defined by

$$T\,(x_1, x_2) = (u_1, u_2) = \left(\ln \frac{x_1}{\overline{x}_1}, \ln \frac{x_2}{\overline{x}_2}\right), \qquad (x_1, x_2) \in X,$$

and whose inverse mapping $T^{-1} : \mathbb{R}^2 \to X$ is given by

$$T^{-1}\,(u_1, u_2) = (x_1, x_2) = \left(\overline{x}_1\, \mathrm{e}^{u_1}, \overline{x}_2\, \mathrm{e}^{u_2}\right), \qquad (u_1, u_2) \in \mathbb{R}^2.$$

The system (1.14) is then transformed into

$$\begin{aligned} \dot{u}_1 &= -c\,(1 - \mathrm{e}^{u_2}), \\ \dot{u}_2 &= r\,(1 - \mathrm{e}^{u_1}), \qquad (u_1, u_2) \in \mathbb{R}^2. \end{aligned} \tag{1.15}$$

Further it follows that $T\,(\overline{x}_1, \overline{x}_2) = (0, 0)$. If $(x_1(t), x_2(t))$, $t \in \mathbb{R}$, is a solution of (1.14), then

$$u_1(t) = \ln \frac{x_1(t)}{\overline{x}_1}, \qquad u_2(t) = \ln \frac{x_2(t)}{\overline{x}_2}$$

is a solution of (1.15). If $(u_1(t), u_2(t))$ is a solution of (1.15), then

$$x_1(t) = \overline{x}_1\, \mathrm{e}^{u_1(t)}, \qquad x_2(t) = \overline{x}_2\, \mathrm{e}^{u_2(t)}$$

is a solution of (1.14).

If $\gamma_{\mathbb{R}}(x_1, x_2)$ is an orbit of the system (1.14) through $(x_1, x_2) \in X$, then $T(\gamma_{\mathbb{R}}(x_1, x_2))$ is an orbit of the system (1.15) through $(u_1, u_2) = T(x_1, x_2)$.

If we can show that every orbit of the system (1.15) is positively compact, then this is also true for every orbit of the system (1.14).

We consider at first a so called Lyapunov function of the form

$$V(u_1, u_2) = r(\mathrm{e}^{u_1} - u_1) + c(\mathrm{e}^{u_2} - u_2) - r - c, \qquad (u_1, u_2) \in \mathbb{R}^2.$$

For this we have

$$V(0,0) = 0 \qquad \text{and} \qquad V(u_1, u_2) > 0 \quad \text{for all} \quad (u_1, u_2) \neq (0,0)$$

and for every $(u_1(t), u_2(t))$ of (1.15) it follows that

$$\frac{\mathrm{d}}{\mathrm{d}t} V(u_1(t), u_2(t)) = r\left(\mathrm{e}^{u_1(t)} - 1\right)\dot{u}_1 + c\left(\mathrm{e}^{u_2(t)} - 1\right)\dot{u}_2 = 0$$

for all $t \in (-\alpha, \alpha) =$ definition interval of $(u_1(t), u_2(t))$.

This further implies that

$$V(u_1(t), u_2(t)) = r\left(\mathrm{e}^{u_1(t)} - u_1(t)\right) + c\left(\mathrm{e}^{u_2(t)} - u_2(t)\right) - r - c = k \quad (1.16)$$

for all $t \in (-\alpha, \alpha)$ where $k \geq 0$ is a constant.

This equality shows that the orbit $\gamma_{\mathbb{R}}(u_1(0), u_2(0))$ is bounded and hence positively compact. This also shows $\alpha = \infty$ (see above!). Hence also the inverse image $T^{-1}(\gamma_{\mathbb{R}}(u_1(0), u_2(0))) = \gamma_{\mathbb{R}}\left(\overline{x}_1\, \mathrm{e}^{u_1(0)}, \overline{x}_2\, \mathrm{e}^{u_2(0)}\right)$ is positively compact.

In order to apply Proposition 1.10 we have to show that in case $(x_1, x_2) \neq (\overline{x}_1, \overline{x}_2)$ the limit set $\Omega(x_1, x_2)$ which belongs to $\gamma_{\mathbb{R}}(x_1, x_2)$ does not contain the only rest point $(\overline{x}_1, \overline{x}_2)$ of the system (1.14). For this it suffices to see that in the case $(u_1(0), u_2(0)) \neq (0,0)$ the limit set $\Omega(u_1(0), u_2(0))$ which belongs to $\gamma_{\mathbb{R}}(u_1(0), u_2(0))$ does not contain the point $(0,0)$. For that purpose we consider a limit point $(u_1, u_2) \in \Omega(u_1(0), u_2(0))$. Then by Lemma 1.1 there exists a sequence $(t_n)_{n\in\mathbb{N}_0}$ with $t_n \to \infty$ and

$$u_1 = \lim_{n\to\infty} u_1(t_n), \qquad u_2 = \lim_{n\to\infty} u_2(t_n)$$

where $(u_1(t), u_2(t))$ is a solution of (1.15). From (1.16) it then follows that

$$V(u_1, u_2) = k = V(u_1(0), u_2(0)) > 0$$

which implies $(u_1, u_2) \neq (0,0)$. By Proposition 1.10 we therefore obtain the

Result. *The omega limit set $\Omega(x_1, x_2)$ of an arbitrary orbit $\gamma_{\mathbb{R}}(x_1, x_2)$ through $(x_1, x_2) \in X$ with $(x_1, x_2) \neq (\overline{x}_1, \overline{x}_2)$ is equal to $\gamma_+(x_1, x_2)$ and simultaneously a periodic orbit.*

1.2.3 Stability: The Direct Method of Lyapunov

As in Section 1.2.1 we at first consider a dynamical system π on a metric space (X, d) with $I = \mathbb{R}$.

Definition. *A rest point $x \in X$ of the flow $\pi : X \times \mathbb{R} \to X$ is called stable, if for every $\varepsilon > 0$ there exists a $\delta = \delta(\varepsilon)$ such that*

$$d(x, y) \leq \delta \quad \Longrightarrow \quad d(x, \pi(x, t)) \leq \varepsilon \qquad \text{for all} \quad t \geq 0,$$

and asymptotically stable, if x is stable and there exists a $\delta_0 > 0$ such that

$$\lim_{t \to \infty} \pi(y, t) = x \qquad \text{for all} \quad y \in X \quad \text{with} \quad d(x, y) \leq \delta_0.$$

In words stability means that a positive semi-orbit $\gamma_+(y)$ stays arbitrarily close to x, if y is sufficiently close to x. Asymptotic stability means that in addition there exists a neighborhood $\{y \in X \mid d(x, y) \leq \delta_0\} = V_{\delta_0}(x)$ such that for every $y \in V_{\delta_0}(x)$ the positive semi-orbit $\gamma_+(y)$ gets arbitrarily close to the rest point x.

After this definition we turn to the first example of a time-continuous dynamical system in Section 1.1 and consider again an autonomous system of the form

$$\dot{x} = f(x) \tag{1.17}$$

where $f \in C^1(W, \mathbb{R}^n)$, $W \subseteq \mathbb{R}^n$ is open and connected. For every $x \in W$ then there exists exactly one solution $\varphi = \varphi(x, t)$, $t \in (-\alpha, \alpha)$ for some $\alpha > 0$ of (1.17) with $\varphi \in C^1((-\alpha, \alpha), W)$ and

$$\varphi(x, 0) = x.$$

We assume that $\alpha = \infty$.

If we define $\pi : X \times \mathbb{R} \to X$, $X = W$, by

$$\pi(x, t) = \varphi(x, t), \qquad x \in X, \; t \in \mathbb{R},$$

then we obtain a flow as being shown in Section 1.1.

A point $x \in X$ is a rest point of this flow, if and only if $f(\overline{x}) = O_n =$ null-vector of $\mathbb{R}^n$.

In order to derive sufficient conditions for the stability and asymptotic stability of such a rest point we will apply the so called direct method of Lyapunov. For that purpose we start with the following

Definition. *A function $V \in C^1(X)$ is called a Lyapunov function with respect to f, if*

$$\dot{V}(x) = \text{grad } V(x)^T f(x) \leq 0 \qquad \text{for all} \quad x \in X.$$

With this definition we formulate the

Proposition 1.12. *If there exists a Lyapunov function $V \in \mathcal{C}^1(X)$ with respect to f which is positive definite with respect to some rest point $x \in X$, i.e., which satisfies the condition*

$$V(\overline{x}) = 0 \quad \textit{and} \quad V(x) > 0 \qquad \textit{for all} \quad x \in X,\ x \neq \overline{x},$$

then $\overline{x}$ is stable.

If in addition

$$\dot{V}(\overline{x}) = 0 \quad \textit{and} \quad \dot{V}(x) < 0 \qquad \textit{for all} \quad x \in X,\ x \neq \overline{x},$$

then $\overline{x}$ is asymptotically stable.

Proof. Since X is open, there is some $r > 0$ such that $B_r(\overline{x}) \subseteq X$ where $B_r(\overline{x})$ denotes the ball around $\overline{x}$ with radius r. Now let $\varepsilon \in (0, r)$ be chosen arbitrarily.

Then

$$k_\varepsilon = \min\{V(x) \mid ||x - \overline{x}||_2 = \varepsilon\} > 0,$$

and because of the continuity of V there is some $\delta = \delta(\varepsilon)$ with $0 < \delta < \varepsilon$ and

$$V(x) < k_\varepsilon \qquad \text{for} \quad ||x - \overline{x}||_2 \leq \delta.$$

Now let $y \in B_\delta(\overline{x})$. Then it follows for the solution $x = x(t)$ of (1.17) with $x(0) = y$

$$\dot{V}(x(t)) = \text{grad } V(x(t))^T f(x(t)) = \text{grad } V(x(t))^T \dot{x}(t) = \frac{\mathrm{d}V}{\mathrm{d}t}(x(t))$$

and hence

$$V(x(t)) - V(y) = \int_0^t \frac{\mathrm{d}V}{\mathrm{d}t}(x(\tau))\ \mathrm{d}\tau = \int_0^t \dot{V}(x(\tau))\ \mathrm{d}\tau \leq 0$$

for every $t \geq 0$. Therefore we have

$$V(x(t)) \leq V(y) < k_\varepsilon \qquad \text{for all} \quad t \geq 0$$

which implies $x(t) \in B_\varepsilon(\overline{x})$ for all $t \geq 0$. If this were not the case, i.e., $x(t) \notin B_\varepsilon(\overline{x})$ or $||\overline{x} - x(t)||_2 > \varepsilon$ for some $t > 0$, then by the intermediate value theorem for continuous functions there exists some $t^* \in (0, t)$ with $||\overline{x} - x(t^*)||_2 = \varepsilon$ which implies $V(x(t^*)) \geq k_\varepsilon$, a contradiction to $V(x(t^*)) < k_\varepsilon$.

Therefore

$$||\overline{x} - \pi(y, t)||_2 \leq \varepsilon \qquad \text{for all} \quad t \geq 0 \quad \text{and all} \quad y \in B_\delta(\overline{x})$$

which shows the stability of $\overline{x}$.

The stability of $\overline{x}$ implies in particular the existence of $b_0 > 0$ and $h > 0$ such that for the solution $x = x(t)$ of (1.17) with $x(0) = x_0 \in X$ it follows that

$$||\overline{x} - x(t)||_2 < h \qquad \text{for all} \quad t \geq 0, \text{ if } \quad ||x_0 - \overline{x}||_2 < b_0.$$

Also, for every sufficiently small $\varepsilon > 0$, there exists some $\delta = \delta(\varepsilon)$ with

$$||\overline{x} - x(t)||_2 < \varepsilon \qquad \text{for all} \quad t \geq 0, \text{ if } \quad ||x_0 - \overline{x}||_2 < \delta.$$

In order to show asymptotic stability it suffices to show the existence of some $T_\varepsilon > 0$ with

$$||\overline{x} - x(t)||_2 < \varepsilon \qquad \text{for all} \quad t \geq T_\varepsilon, \text{ if } \quad ||x_0 - \overline{x}||_2 < b_0.$$

Assumption: There exists some $x_0 \in X$ with

$$||x_0 - \overline{x}||_2 < b_0 \qquad \text{and} \qquad ||\overline{x} - x(t)||_2 \geq \delta(\varepsilon) \quad \text{for some} \quad t \geq 0.$$

Let $\gamma > 0$ be chosen such that

$$\dot{V}(x(t)) < -\gamma, \quad \text{if} \quad \delta(\varepsilon) \leq ||\overline{x} - x(t)||_2 < h.$$

Then it follows that $\mathrm{d}V/\mathrm{d}t\,(x(t)) < -\gamma$ and hence

$$V(x(t)) < V(x_0) - \gamma t, \quad \text{if} \quad \delta(\varepsilon) \leq ||\overline{x} - x(t)||_2 < h.$$

Let $\beta, k > 0$ be chosen such that

$$\beta \leq V(x) \leq k, \quad \text{if} \quad \delta(\varepsilon) \leq ||\overline{x} - x||_2 < h.$$

Then we put $T_\varepsilon = (k - \beta)/\gamma$ and conclude for $t > T_\varepsilon$ that

$$V(x(t)) < V(x_0) - \gamma T_\varepsilon \leq k - (k - \beta) = \beta.$$

Therefore there must exist some $t_0 \in [0, T_\varepsilon]$ with $||\overline{x} - x(t_0)||_2 < \delta(\varepsilon)$.

The stability of $\overline{x}$ then implies

$$||\overline{x} - x(t)||_2 < \varepsilon \qquad \text{for all} \quad t \geq t_0, \quad \text{hence for all} \quad t \geq T_\varepsilon.$$

This completes the proof. □

Let us demonstrate Proposition 1.12 by two examples. At first we consider the system

$$\begin{aligned} \dot{u}_1(t) &= -c\left(1 - \mathrm{e}^{u_2(t)}\right), \\ \dot{u}_2(t) &= r\left(1 - \mathrm{e}^{u_1(t)}\right), \qquad t \in \mathbb{R}. \end{aligned} \tag{1.15}$$

The only rest point of this system is $(0, 0)$.

If we define $V : X = \mathbb{R}^2 \to \mathbb{R}$ by

$$V(u_1, u_2) = r(e^{u_1} - u_1) + c(e^{u_2} - u_2) - r - c$$

for $(u_1, u_2) \in \mathbb{R}^2$, then it follows that

$$V(0,0) = 0 \qquad \text{and} \qquad V(u_1, u_2) > 0 \quad \text{for all} \quad (u_1, u_2) \neq (0,0).$$

The function $V = V(u_1, u_2)$ is obviously in $\mathcal{C}^1(\mathbb{R}^2)$ and positive definite with respect to $(0,0)$. Further it follows that

$$\dot{V}(u_1, u_2) = -rc(e^{u_1} - 1)(1 - e^{u_2}) + rc(e^{u_2} - 1)(1 - e^{u_1}) = 0$$

for all $(u_1, u_2) \in X$, i.e., V is a Lyapunov function. By Proposition 1.12 therefore $(0,0)$ is a stable rest point.

As a second example we choose a modification of the system (1.14) and consider the system

$$\begin{aligned} \dot{x}_1(t) &= -c\, x_1(t) - b\, x_1^2(t) + a\, x_1(t)x_2(t), \\ \dot{x}_2(t) &= r\, x_2(t) - k\, x_2^2(t) - p\, x_1(t)x_2(t), \qquad t \in \mathbb{R}, \end{aligned} \tag{1.18}$$

where $a, b, c, k, p, r \in \mathbb{R}$ are positive constants. We again choose

$$X = \left\{(x_1, x_2) \in \mathbb{R}^2 \mid x_1 > 0,\ x_2 > 0\right\}.$$

If we require that $ra - kc > 0$ then the only rest point $(\overline{x}_1, \overline{x}_2) \in X$ of the system (1.18) is given by

$$\overline{x}_1 = \frac{ra - kc}{bk + ap}, \qquad \overline{x}_2 = \frac{br + pc}{bk + ap}.$$

If we again apply the homeomorphic mapping $T : X \to \mathbb{R}^2$ which is given by

$$T(x_1, x_2) = (u_1, u_2) = \left(\ln \frac{x_1}{\overline{x}_1}, \ln \frac{x_2}{\overline{x}_2}\right), \qquad (x_1, x_2) \in X,$$

then the system (1.18) is transformed into the system

$$\begin{aligned} \dot{u}_1(t) &= b\, \overline{x}_1 \left(1 - e^{u_1(t)}\right) - a\, \overline{x}_2 \left(1 - e^{u_2(t)}\right), \\ \dot{u}_2(t) &= p\, \overline{x}_1 \left(1 - e^{u_1(t)}\right) + k\, \overline{x}_2 \left(1 - e^{u_2(t)}\right), \qquad t \in \mathbb{R}, \end{aligned} \tag{1.19}$$

and the only rest point of system (1.19) is given by $(0,0) = T(\overline{x}_1, \overline{x}_2)$.

If $(u_1(t), u_2(t))$, $t \in (-\alpha, \alpha)$ for some $\alpha > 0$, is a solution of (1.19), then $(x_1(t), x_2(t)) = \left(\overline{x}_1\, e^{u_1(t)}, \overline{x}_2\, e^{u_2(t)}\right)$, $t \in (-\alpha, \alpha)$, is a solution of (1.18) and if this is the case, then $(u_1(t), u_2(t)) = (\ln(x_1(t)/\overline{x}_1), \ln(x_2(t)/\overline{x}_2))$ is a solution of (1.19).

If we define a function $V : \mathbb{R}^2 \to \mathbb{R}$ by

$$V(u_1, u_2) = p\,\overline{x}_1\,(\mathrm{e}^{u_1} - u_1) + a\,\overline{x}_2\,(\mathrm{e}^{u_2} - u_2) - p\,\overline{x}_1 - a\,\overline{x}_2$$

for $(u_1, u_2) \in \mathbb{R}^2$, then $V \in \mathcal{C}^1(\mathbb{R}^2)$ and

$$V(0,0) = 0 \qquad \text{and} \qquad V(u_1, u_2) > 0 \quad \text{for all} \quad (u_1, u_2) \neq (0,0).$$

Hence V is positive definite with respect to $(0,0)$. Further it follows that

$$\dot{V}(u_1, u_2) = -pb\,\overline{x}_1^2\,(1 - \mathrm{e}^{u_1})^2 - ak\,\overline{x}_2^2\,(1 - \mathrm{e}^{u_2})^2 \leq 0 \tag{1.20}$$

for all $(u_1, u_2) \in \mathbb{R}^2$.

Now let $(u_1(t), u_2(t))\,,\ t \in (-\alpha, \alpha)$ for some $\alpha > 0$ a solution of (1.19). Then for every $t \in \mathbb{R}$ it follows that

$$\dot{V}(u_1(t), u_2(t)) = \operatorname{grad} V(u_1(t), u_2(t))^T \begin{pmatrix} \dot{u}_1(t) \\ \dot{u}_2(t) \end{pmatrix} = \frac{\mathrm{d}}{\mathrm{d}t}\, V(u_1(t), u_2(t)) \leq 0,$$

hence

$$V(u_1(t), u_2(t)) \leq V(u_1(0), u_2(0)) \qquad \text{for all} \quad t \in \mathbb{R}$$

and

$$\{(u_1(t), u_2(t))\,,\ t \in \mathbb{R}\} \subseteq V^{-1}\left[0, V(u_1(0), u_2(0))\right]$$

which shows that $(u_1(t), u_2(t))$ is defined on all of $\mathbb{R}$.

The inequality (1.20) shows that $V : \mathbb{R}^2 \to \mathbb{R}$ is a Lyapunov function. From Proposition 1.12 it therefore follows that $(0,0)$ is a stable rest point.

In addition it follows from (1.20) that

$$\dot{V}(0,0) = 0 \qquad \text{and} \qquad \dot{V}(u_1, u_2) < 0 \quad \text{for all} \quad (u_1, u_2) \neq (0,0)$$

which implies that $(0,0)$ is an asymptotically stable rest point.

After these two examples we consider a linear system

$$\dot{x} = A\,x, \qquad x \in \mathbb{R}^n, \tag{1.21}$$

where A is a real $n \times n-$matrix. Then for every $x \in \mathbb{R}^n$ there exists a unique solution $\varphi \in \mathcal{C}^1(\mathbb{R}, \mathbb{R}^n)$ with $\varphi(0) = x$ and by the definition

$$\pi(x,t) = \varphi(x,t) \qquad \text{for all} \quad x \in X = \mathbb{R}^n,\ t \in I = \mathbb{R},$$

we obtain a dynamical system $\pi : X \times I \to X$.

We assume that all the eigenvalues of A have negative real parts which implies that A is non-singular and $x = \Theta_n$ is the only rest point of the system (1.21).

Now let C be an arbitrary symmetric and positive definite matrix. Then we define the symmetric and positive definite matrix

$$B = \int_0^{\infty} \mathrm{e}^{tA^T} C \, \mathrm{e}^{tA} \, \mathrm{d}t$$

(B is well defined, since there exist positive constants K and α with $||\mathrm{e}^{tA}|| \leq K \, \mathrm{e}^{-\alpha t}, \ t \geq 0$). If we define

$$V(x) = x^T B \, x, \qquad x \in \mathbb{R}^n,$$

it follows that $V : \mathbb{R}^n \to \mathbb{R}_+$ is positive definite with respect to Θ_n.

Further it follows that

$$\dot{V}(x) = 2 \, x^T BA \, x = x^T BA \, x + x^T A^T B \, x = x^T \left(BA + A^T B \right) \, x < 0$$

for all $x \in \mathbb{R}^n$ and $\dot{V}(\Theta_n) = 0$ because of

$$A^T B + BA = \int_0^{\infty} \frac{\mathrm{d}}{\mathrm{d}t} \left(\mathrm{e}^{tA^T} C \, \mathrm{e}^{tA} \right) \, \mathrm{d}t = -C.$$

Hence V is a Lyapunov function and by Proposition 1.12 it follows that Θ_n is an asymptotically stable rest point of the system (1.21).

The asymptotic stability of Θ_n also implies that the eigenvalues of the matrix A have only negative real parts (see, for instance, H.W. Knobloch and F. Kappel: Gewönliche Differentialgleichungen, Verlag B.G. Teubner, Stuttgart 1974, Kap. III, Satz 7.5).

Let us return to the system

$$\dot{x} = f(x) \tag{1.17}$$

with $f \in \mathcal{C}^1(W, \mathbb{R}^n)$ where $W \subseteq \mathbb{R}^n$ is open and connected. Now let $\overline{x} \in W$ be a rest point of this system, i.e., $f(\overline{x}) = \Theta_n$. Then for every $x \in W$ there is a representation

$$f(x) = A(x - \overline{x}) + r(x - \overline{x})$$

where $A = \left(\dfrac{\partial f_i(\overline{x})}{\partial x_j} \right)_{1 \leq i,j \leq n}$ is the Jacobi matrix of f in $\overline{x}$ and

$$r : (W - \overline{x}) \to \mathbb{R}^n$$

a vector function with

$$\lim_{||x - \overline{x}||_2 \to 0} \frac{||r(x - \overline{x})||_2}{||x - \overline{x}||_2} = 0.$$

We assume that A only possesses eigenvalues with negative real parts.

According to the considerations above then there exists a symmetric and positive definite matrix B such that

$$A^T B + BA = -I \qquad \text{where} \quad I = n \times n - \text{unit matrix.}$$

If we define

$$V(x) = (x - \overline{x})^T B \; (x - \overline{x}) \qquad \text{for} \quad x \in W,$$

then V is positive definite with respect to $\overline{x}$. Further it follows that

$$\begin{aligned} \dot{V}(x) &= (x - \overline{x})^T B \; f(x) + f(x)^T B \, (x - \overline{x}) \\ &= (x - \overline{x})^T \left(BA + A^T B \right) \; (x - \overline{x}) + 2 \, (x - \overline{x})^T \; B \; r \, (x - \overline{x}) \\ &\le - ||x - \overline{x}||_2^2 + 2 ||x - \overline{x}||_2 \; ||B|| \; ||r \, (x - \overline{x}) \, ||_2 \\ &= - ||x - \overline{x}||_2^2 \, (1 - 2 \; ||B|| \; \mathcal{E} \, (||x - \overline{x}||_2)) \end{aligned}$$

where

$$\mathcal{E} \, (||x - \overline{x}||_2) = \frac{||r \, (x - \overline{x}) \, ||_2}{||x - \overline{x}||_2} \to 0 \qquad \text{for} \quad ||x - \overline{x}||_2 \to 0.$$

Let $r > 0$ be such that

$$\mathcal{E} \, (||x - \overline{x}||_2) \le \frac{1}{4 \; ||B||} \qquad \text{for all} \quad x \in \mathbb{R}^n \quad \text{with} \quad ||x - \overline{x}||_2 \le r$$

and $B_r \, (\overline{x}) = \{ x \in \mathbb{R}^n \mid ||x - \overline{x}||_2 \le r \} \subseteq W$. Then it follows that

$$\dot{V} \, (\overline{x}) = 0 \quad \text{and} \quad \dot{V}(x) < 0 \qquad \text{for all} \quad x \in B_r \, (\overline{x}) \, .$$

As a consequence of the proof of Proposition 1.12 we therefore obtain

Proposition 1.13. *Let $\overline{x} \in W$ be a rest point of the system (1.17) such that the Jacobi matrix $\left(\frac{\partial f_i}{\partial x_j} \, (\overline{x}) \right)_{1 \le i,j \le n}$ has only eigenvalues with negative real parts. Then $\overline{x}$ is asymptotically stable.*

Let us demonstrate this result by the system

$$\begin{aligned} \dot{u}_1 &= -a \; \overline{x}_2 \, (1 - \mathrm{e}^{u_2}) \, , \\ \dot{u}_2 &= p \; \overline{x}_1 \, (1 - \mathrm{e}^{u_1}) + k \; \overline{x}_2 \, (1 - \mathrm{e}^{u_2}) \, , \qquad (u_1, u_2) \in \mathbb{R}^2 \end{aligned} \tag{1.22}$$

where

$$\overline{x}_1 = \frac{1}{p} \left(r - k \frac{c}{a} \right), \qquad \overline{x}_2 = \frac{c}{a},$$

$a, c, k, p, r > 0$ and $ra - kc > 0$.

The point $(0, 0)$ is the only rest point of this system.

The Jacobi matrix of the right-hand side of (1.22) is given by

$$\begin{pmatrix} 0 & a\,\overline{x}_2 e^{u_2} \\ -p\,\overline{x}_1 e^{u_1} & -k\,\overline{x}_2 e^{u_2} \end{pmatrix}, \qquad (u_1, u_2) \in \mathbb{R}^2.$$

For $u_1 = 0,\ u_2 = 0$ we obtain the matrix

$$A = \begin{pmatrix} 0 & a\,\overline{x}_2 \\ -p\,\overline{x}_1 & -k\,\overline{x}_2 \end{pmatrix}$$

with the eigenvalues

$$\lambda_{1,2} = -\frac{k\,\overline{x}_2}{2} \pm \sqrt{\frac{k^2\,\overline{x}_2^2}{4} - ap\,\overline{x}_1\overline{x}_2}$$

which are either both real and negative or have a negative real part.

By Proposition 1.13 the point $(0,0)$ is an asymptotically stable rest point of the system (1.22).

1.2.4 Application to a General Predator-Prey Model

We consider n populations $X_i,\ i = 1,\dots,n$, of animals or plants that are living in mutual predator-prey relations or are pairwise neutral to each other. Let us denote by $x_i(t)$ the size of population X_i at the time t. We assume the temporal development of these sizes to be described by the following system of differential equations

$$\dot{x}_i(t) = \left(c_i + \sum_{j=1}^{n} c_{ij}\, x_j(t) \right) x_i(t) \tag{1.23}$$
$$\text{for} \quad i = 1,\dots,n$$

where

$$c_{ii} \le 0 \qquad \text{for} \quad i = 1,\dots,n$$

and, for $i \neq j$,

$$\begin{aligned} &c_{ij} > 0, \quad \text{if} \quad X_i = \textit{predator} \text{ and } X_j = \textit{prey}, \\ &c_{ij} < 0, \quad \text{if} \quad X_i = \textit{prey} \text{ and } X_j = \textit{predator}, \\ &c_{ij} = c_{ji} = 0, \quad \text{if} \quad X_i \text{ and } X_j \text{ are neutral to each other.} \end{aligned}$$

This model is a generalization of the predator-prey models for two populations which were considered in Sections 1.2.2 and 1.2.3.

Let us further assume that there is a rest point or an equilibrium state $\overline{x} = (\overline{x}_1, \ldots, \overline{x}_n)$ of (1.23), i.e. a solution of the linear system

$$\sum_{j=1}^{n} c_{ij}\,\overline{x}_j = -c_i, \qquad i = 1, \ldots, n, \tag{1.24}$$

with

$$\overline{x}_j > 0, \qquad j = 1, \ldots, n. \tag{1.25}$$

Then (1.23) can be written in the form

$$\dot{x}_i(t) = \left(\sum_{j=1}^{n} c_{ij}\,(x_j(t) - \overline{x}_j) \right) x_i(t) \tag{1.26}$$

$$\text{for} \quad i = 1, \ldots, n.$$

We assume that, for every initial state $x^0 \in \mathbb{R}^n$ with $x_i^0 > 0$ for all $i = 1, \ldots, n$, there is exactly one solution $x(t) = (x_1(t), \ldots, x_n(t))$ of (1.26) with

$$x_i(0) = x_i^0, \quad x_i(t) > 0 \qquad \text{for all} \quad t > 0 \quad \text{and} \quad i = 1, \ldots, n.$$

The system (1.26) has the form (1.17) with

$$f(x) = \begin{pmatrix} \sum\limits_{j=1}^{n} c_{1j}\,(x_j - \overline{x}_j)\,x_1 \\ \vdots \\ \sum\limits_{j=1}^{n} c_{nj}\,(x_j - \overline{x}_j)\,x_n \end{pmatrix}, \qquad x \in \mathbb{R}^n,$$

and the Jacobi matrix of f in $\overline{x}$ is given by

$$J_f\,(\overline{x}) = \begin{pmatrix} c_{11}\,\overline{x}_1 & \ldots & c_{1n}\,\overline{x}_1 \\ \vdots & & \vdots \\ c_{n1}\,\overline{x}_n & \ldots & c_{nn}\,\overline{x}_n \end{pmatrix}. \tag{1.27}$$

By Proposition 1.13 the equilibrium state $\overline{x}$ is asymptotically stable, if the Jacobi matrix (1.27) has only eigenvalues with negative real parts.

Let us consider the case $n = 2$ where X_1 is the population of predators and X_2 that of preys. The system (1.23) reads

$$\begin{aligned} \dot{x}_1(t) &= (c_1 + c_{11}\,x_1(t) + c_{12}\,x_2(t))\,x_1(t), \\ x_2(t) &= (c_2 + c_{21}\,x_1(t) + c_{22}\,x_2(t))\,x_2(t) \end{aligned} \tag{1.23}$$

with $c_{11} \leq 0,\ c_{22} \leq 0,\ c_{12} > 0,\ c_{21} < 0$. It is reasonable to assume that $c_1 < 0$ and $c_2 > 0$.

The system (1.24) reads

$$c_{11}\,\overline{x}_1 + c_{12}\,\overline{x}_2 = -c_1,$$
$$c_{21}\,\overline{x}_1 + c_{22}\,\overline{x}_2 = -c_2$$

and has the unique solution

$$\overline{x}_1 = \frac{-c_1\,c_{22} + c_2\,c_{12}}{c_{11}\,c_{22} - c_{12}\,c_{21}}, \qquad \overline{x}_2 = \frac{c_1\,c_{21} - c_2\,c_{11}}{c_{11}\,c_{22} - c_{12}\,c_{21}}. \tag{1.28}$$

This implies that $\overline{x}_1 > 0$ and $\overline{x}_2 > 0$, if and only if the condition

$$-c_1\,c_{22} + c_2\,c_{12} > 0 \tag{1.29}$$

is satisfied. This is for instance the case, if $c_{22} = 0$.

The Jacobi matrix

$$J_f\,(\overline{x}) = \begin{pmatrix} c_{11}\,\overline{x}_1 & c_{12}\,\overline{x}_1 \\ c_{21}\,\overline{x}_2 & c_{22}\,\overline{x}_2 \end{pmatrix}$$

of f at $\overline{x} = (\overline{x}_1, \overline{x}_2)$ has the eigenvalues

$$\lambda_{1,2} = \frac{c_{11}\,\overline{x}_1 + c_{22}\,\overline{x}_2}{2} \pm \sqrt{\frac{(c_{11}\,\overline{x}_1 + c_{22}\,\overline{x}_2)^2}{4} - (c_{11}\,c_{22} - c_{12}\,c_{21})\;\overline{x}_1\overline{x}_2}.$$

This implies $\Re\,(\lambda_{1,2}) < 0$, if $c_{11} + c_{22} < 0$.

Proposition 1.13 therefore leads to the

Result. *If the conditions $c_{11} + c_{22} < 0$ and (1.29) are satisfied, then the system (1.23) has exactly one equilibrium state $\overline{x} = (\overline{x}_1, \overline{x}_2)$ with $\overline{x}_1 > 0$ and $\overline{x}_2 > 0$ which is given by (1.28) and which is asymptotically stable.*

In Section 1.2.3 we have obtained the same result under the assumption $c_{11} < 0$ and $c_{22} < 0$ by applying Lyapunov's method. We also considered the case $c_{11} = c_{22} = 0$ in which it only follows that $\overline{x} = (\overline{x}_1, \overline{x}_2)$ is a stable equilibrium state.

Next we consider the case $n = 3$ where we assume that X_1 and X_2 are predator populations that have the X_3–population as prey and are neutral to each other. This leads to

$$c_{12} = 0, \quad c_{13} > 0, \quad c_{23} > 0,$$
$$c_{21} = 0, \quad c_{31} < 0, \quad c_{32} < 0.$$

In addition we assume

$$c_{11} < 0, \quad c_{22} < 0, \quad c_{33} < 0.$$

Finally it is reasonable to assume that

$$c_1 < 0, \quad c_2 < 0, \quad c_3 > 0.$$

The system (1.24) reads

$$\begin{aligned} c_{11}\,\overline{x}_1 + c_{12}\,\overline{x}_2 + c_{13}\,\overline{x}_3 &= -c_1, \\ c_{21}\,\overline{x}_1 + c_{22}\,\overline{x}_2 + c_{23}\,\overline{x}_3 &= -c_2, \\ c_{31}\,\overline{x}_1 + c_{32}\,\overline{x}_2 + c_{33}\,\overline{x}_3 &= -c_3. \end{aligned}$$

and has the unique solution

$$\begin{aligned} \overline{x}_1 &= \frac{(-c_{23}\,c_{32} + c_{22}\,c_{33})\,c_1 + c_{13}\,c_{32}\,c_2 - c_{22}\,c_{13}\,c_3}{c_{11}\,c_{23}\,c_{32} - c_{22}\,c_{33}\,c_{11} + c_{22}\,c_{13}\,c_{31}}, \\ \overline{x}_2 &= \frac{(c_{33}\,c_{11} - c_{13}\,c_{31})\,c_2 - c_{23}\,c_{11}\,c_3 + c_{23}\,c_{31}\,c_1}{c_{11}\,c_{23}\,c_{32} - c_{22}\,c_{33}\,c_{11} + c_{22}\,c_{13}\,c_{31}}, \\ \overline{x}_3 &= \frac{-c_2\,c_{32}\,c_{11} + c_3\,c_{22}\,c_{11} - c_1\,c_{22}\,c_{31}}{c_{11}\,c_{23}\,c_{32} - c_{22}\,c_{33}\,c_{11} + c_{22}\,c_{13}\,c_{31}} > 0. \end{aligned} \tag{1.30}$$

This implies that $\overline{x}_i > 0$ for $i = 1, 2, 3$, if and only if the conditions

$$\begin{aligned} (c_{22}\,c_{33} - c_{23}\,c_{32})\,c_1 &> c_{22}\,c_{13}\,c_3 - c_{13}\,c_{32}\,c_2, \\ (c_{11}\,c_{33} - c_{13}\,c_{31})\,c_2 &> c_{11}\,c_{23}\,c_3 - c_{23}\,c_{31}\,c_1 \end{aligned} \tag{1.31}$$

are satisfied.

The characteristic polynomial of $J_f(\overline{x}_1, \overline{x}_2, \overline{x}_3)$ is given by

$$P(\lambda) = -\lambda^3 - a_1\,\lambda^2 - a_2\,\lambda - a_3, \qquad \lambda \in \mathbb{C},$$

where

$$\begin{aligned} a_1 &= -c_{11}\,\overline{x}_1 - c_{22}\,\overline{x}_2 - c_{33}\,\overline{x}_3 > 0, \\ a_2 &= c_{11}\,c_{22}\,\overline{x}_1\overline{x}_2 + (c_{22}\,c_{33} - c_{23}\,c_{31})\,\overline{x}_2\overline{x}_3 + (c_{11}\,c_{33} - c_{13}\,c_{31})\,\overline{x}_1\overline{x}_3, \\ a_3 &= \underbrace{(-c_{11}\,c_{22}\,c_{33} + c_{11}\,c_{23}\,c_{32} + c_{22}\,c_{13}\,c_{31})\,\overline{x}_1\overline{x}_2\overline{x}_3}_{-\det J_f(\overline{x}_1, \overline{x}_2, \overline{x}_3)} > 0. \end{aligned}$$

By a theorem of Hurwitz (see [9]) all the eigenvalues of $J_f(\overline{x}_1, \overline{x}_2, \overline{x}_3)$ which are zeros of $P = P(\lambda)$ have a negative real part, if and only if

$$a_1\,a_2 - a_3 > 0$$

which can be verified in this case.

Result. *There exists exactly one equilibrium state* $(\overline{x}_1, \overline{x}_2, \overline{x}_3)$ *of (1.23) for* $n = 3$ *with* $\overline{x}_i > 0$ *for* $i = 1, 2, 3$ *which is given by (1.30) and is asymptotically stable, if and only if the conditions (1.31) are satisfied.*

The method above for deriving sufficient conditions for asymptotic stability of equilibrium states becomes rather complicated, if $n > 3$. Therefore we will apply Lyapunov's method in order to find sufficient conditions. For this purpose we define a homeomorphic mapping $T : X = \{x \in \mathbb{R}^n \mid x_1 > 0 \text{ for } i = 1, \ldots, n\} \to \mathbb{R}^n$ by

$$T(x_1, \ldots, x_n) = (u_1, \ldots, u_n) = \left(\ln \frac{x_1}{\overline{x}_1}, \ldots, \ln \frac{x_n}{\overline{x}_n}\right), \qquad (x_1, \ldots, x_n) \in X,$$

whose inverse mapping $T^{-1} : \mathbb{R}^n \to X$ is given by

$$T^{-1}(u_1, \ldots, u_n) = (x_1, \ldots, x_n) = (\overline{x}_1 \, \mathrm{e}^{u_1}, \ldots, \overline{x}_n \, \mathrm{e}^{u_n}), \quad (u_1, \ldots, u_n) \in \mathbb{R}^n.$$

By this mapping the system (1.26) is transformed into the system

$$\dot{u}_i(t) = \sum_{j=1}^{n} c_{ij} \, \overline{x}_j \left(\mathrm{e}^{u_j(t)} - 1\right) \quad \text{for} \quad i = 1, \ldots, n. \tag{1.32}$$

This system has $(0, \ldots, 0) = T(\overline{x}_1, \ldots, \overline{x}_n)$ as equilibrium state. If $x \in \mathcal{C}^1(\mathbb{R}_+, X)$ is a solution of the system (1.23), then $u = Tx \in \mathcal{C}^1(\mathbb{R}_+, \mathbb{R}^n)$ is a solution of (1.32) and if $u \in \mathcal{C}^1(\mathbb{R}_+, \mathbb{R}^n)$ is a solution of (1.32), then $x = T^{-1}(u)$ is a solution of (1.23). This implies (according to the assumption above) that for every $u^0 \in \mathbb{R}^n$ there is exactly one solution $u \in \mathcal{C}^1(\mathbb{R}_+, \mathbb{R}^n)$ of (1.32) with $u(0) = u^0$. Further it follows that $x \in X$ is an asymptotically stable equilibrium state of the system (1.23), if and only if $\Theta_n = T(\overline{x})$ is an asymptotically stable equilibrium state of the system (1.32).

Now let us define a function $V : \mathbb{R}^n \to \mathbb{R}$ by

$$V(u_1, \ldots, u_n) = \sum_{i=1}^{n} \sum_{j=1}^{n} c_{ji} \operatorname{sgn}(-c_j) \, \overline{x}_i \, \left(\mathrm{e}^{u_i} - u_i\right) - \sum_{j=1}^{n} |c_j|.$$

Then it follows that

$$\begin{aligned} &V(0, \ldots, 0) = 0 \quad \text{and} \\ &V(u_1, \ldots, u_n) > 0 \quad \text{for all} \quad (u_1, \ldots, u_n) \neq (0, \ldots, 0), \end{aligned}$$

if

$$\sum_{j=1}^{n} c_{ji} \operatorname{sgn}(-c_j) > 0 \qquad \text{for all} \quad i = 1, \ldots, n, \tag{1.33}$$

i.e., V is positive definite with respect to Θ_n.

Further it follows that

$$
\begin{aligned}
&\dot{V}(u_1,\dots,u_n)\\
&\quad = \sum_{i=1}^{n}\left(\sum_{j=1}^{n} c_{ji}\,\mathrm{sgn}\,(-c_j)\,\overline{x}_i\,\left(\mathrm{e}^{u_i}-1\right)\right)\left(\sum_{k=1}^{n} c_{ik}\,\overline{x}_k\,\left(\mathrm{e}^{u_k}-1\right)\right)\\
&\quad = \sum_{i=1}^{n} c_{ii}\left(\sum_{j=1}^{n} c_{ji}\,\mathrm{sgn}\,(-c_j)\right)\overline{x}_i^2\,\left(\mathrm{e}^{u_i}-1\right)^2\\
&\qquad\quad + \sum_{i,k=1}^{n}\left(c_{ik}\left(\sum_{j=1}^{n} c_{ji}\,\mathrm{sgn}\,(-c_j)\right)\right.\\
&\qquad\qquad\qquad \left.+\,c_{ki}\left(\sum_{j=1}^{n} c_{jk}\,\mathrm{sgn}\,(-c_j)\right)\right)\overline{x}_i\overline{x}_k\,\left(\mathrm{e}^{u_i}-1\right)\left(\mathrm{e}^{u_k}-1\right)\\
&\quad = \sum_{i=1}^{n} c_{ii}\left(\sum_{j=1}^{n} c_{ji}\,\mathrm{sgn}\,(-c_j)\right)\overline{x}_i^2\,\left(\mathrm{e}^{u_i}-1\right)^2 < 0
\end{aligned}
$$

for $(u_1,\dots,u_n) \neq (0,\dots,0)$, if

$$c_{ii} < 0 \qquad \text{for all} \quad i = 1,\dots,n \tag{1.34}$$

and

$$c_{ik}\left(\sum_{j=1}^{n} c_{ji}\,\mathrm{sgn}\,(-c_j)\right) + c_{ki}\left(\sum_{j=1}^{n} c_{jk}\,\mathrm{sgn}\,(-c_j)\right) = 0 \tag{1.35}$$

$$\text{for all} \quad i \neq k.$$

Finally we have $\dot{V}(0,\dots,0) = 0$.

By Proposition 1.12 it therefore follows that Θ_n is an asymptotically stable equilibrium state of the system (1.32) and hence $\overline{x} = T^{-1}(\Theta_n)$ an asymptotically stable equilibrium state of the system (1.26), (1.23), if the conditions (1.33), (1.34), (1.35) are satisfied.

If $c_{ii} = 0$ for at least one $i = 1,\dots,n$, then Θ_n is only a stable equilibrium state of the system (1.32).

Next we consider the special cases $n = 2$ and $n = 3$. For $n = 2$ we assume that X_1 is a predator population which has the population X_2 as prey. This leads to

$$c_{12} > 0 \qquad \text{and} \qquad c_{21} < 0.$$

We further assume that $c_{11} = c_{22} = 0$.

This implies

$$\overline{x}_1 = -\frac{c_2}{c_{21}} \qquad \text{and} \qquad \overline{x}_2 = -\frac{c_1}{c_{12}}$$

which requires $c_1 < 0$, $c_2 > 0$ in order to make $\overline{x}_1$ and $\overline{x}_2$ positive. The conditions (1.33) and (1.35) are satisfied which implies that $(0,0)$ is a stable equilibrium state of the system (1.32) for $n = 2$.

For $n = 3$ we consider at first the case that X_1 is a predator population that has the populations X_2 and X_3 as prey and would decay exponentially in the absence of X_2 and X_3. We further assume that X_2 and X_3 would grow logistically in the absence of X_1 and are neutral to each other. All this leads to the conditions

$$\begin{array}{lll} c_1 < 0, & c_2 > 0, & c_3 > 0, \\ c_{11} = 0, & c_{22} < 0, & c_{33} < 0, \\ c_{12} > 0, & c_{13} > 0, & c_{23} = 0, \\ c_{21} < 0, & c_{31} < 0, & c_{32} = 0. \end{array}$$

The condition (1.33) read

$$\begin{aligned} -c_{21} - c_{31} &> 0, \\ c_{12} - c_{22} &> 0, \\ c_{13} - c_{33} &> 0 \end{aligned}$$

and are satisfied.

The conditions (1.35) turn out to be equivalent to

$$c_{12}\, c_{31} + c_{21}\, c_{22} = 0, \qquad c_{13}\, c_{21} + c_{31}\, c_{33} = 0. \tag{1.36}$$

By Proposition 1.12 we conclude that $(0,0,0)$ is a stable equilibrium state of the system (1.32) for $n = 3$, if the conditions (1.36) are satisfied.

The unique solution of the system (1.24) for $n = 3$ is given by

$$\begin{aligned} \overline{x}_1 &= \frac{c_{22}\, c_{33}\, c_1 - c_{12}\, c_{33}\, c_2 - c_{13}\, c_{22}\, c_3}{c_{12}\, c_{21}\, c_{33} + c_{13}\, c_{31}\, c_{22}}, \\ \overline{x}_2 &= \frac{-c_{13}\, c_{31}\, c_2 - c_{21}\, c_{33}\, c_1 + c_{21}\, c_{13}\, c_3}{c_{12}\, c_{21}\, c_{33} + c_{13}\, c_{31}\, c_{22}}, \\ \overline{x}_3 &= \frac{-c_{12}\, c_{21}\, c_3 - c_{31}\, c_{22}\, c_1 + c_{31}\, c_{12}\, c_2}{c_{12}\, c_{21}\, c_{33} + c_{13}\, c_{31}\, c_{22}}, \end{aligned}$$

and $\overline{x}_i > 0$ for $i = 1, 2, 3$ is satisfied, if and only if

$$\begin{aligned} c_{22}\, c_{33}\, c_1 - c_{12}\, c_{33}\, c_2 - c_{13}\, c_{22}\, c_3 &> 0, \\ -c_{13}\, c_{31}\, c_2 - c_{21}\, c_{33}\, c_1 + c_{21}\, c_{13}\, c_3 &> 0, \\ -c_{12}\, c_{21}\, c_3 - c_{31}\, c_{22}\, c_1 + c_{31}\, c_{12}\, c_2 &> 0. \end{aligned}$$

Next we consider the case that X_1 and X_2 are predator populations which have the population X_3 as prey and are neutral to each other. Further we assume that X_1 and X_2 would decay exponentially in the absence of X_3 and that X_3 would increase logistically in the absence of X_1 and X_2. This leads to the conditions

$$\begin{aligned} &c_1 < 0, && c_2 < 0, && c_3 > 0, \\ &c_{11} = 0, && c_{22} = 0, && c_{33} < 0, \\ &c_{12} = 0, && c_{13} > 0, && c_{23} > 0, \\ &c_{21} = 0, && c_{31} < 0, && c_{32} < 0. \end{aligned}$$

The above choice of a Lyapunov function $V : \mathbb{R}^3 \to \mathbb{R}$ does not lead to sufficient conditions for stability of $(0,0,0)$ as we shall see later. We therefore define

$$\begin{aligned} V(u_1,u_2,u_3) &= -c_{31}\,\overline{x}_1\,(\mathrm{e}^{u_1} - u_1) - c_{32}\,\overline{x}_2\,(\mathrm{e}^{u_2} - u_2) \\ &\quad -c_{33}\,\overline{x}_3\,(\mathrm{e}^{u_3} - u_3) - c_3 \\ &\quad \text{for} \quad (u_1,u_2,u_3) \in \mathbb{R}^3 \end{aligned}$$

and conclude that

$$V(0,0,0) = 0, \qquad V(u_1,u_2,u_3) > 0 \quad \text{for all} \quad (u_1,u_2,u_3) \neq (0,0,0).$$

Further we obtain

$$\begin{aligned} &\dot{V}(u_1,u_2,u_3) \\ &\quad = -c_{31}c_{13}\,\overline{x}_1\overline{x}_3\,(\mathrm{e}^{u_1} - 1)(\mathrm{e}^{u_3} - 1) - c_{32}c_{23}\,\overline{x}_2\overline{x}_3\,(\mathrm{e}^{u_2} - 1)(\mathrm{e}^{u_3} - 1) \\ &\qquad -c_{33}\,\overline{x}_3\,(\mathrm{e}^{u_3} - 1)\,\underbrace{(c_3 + c_{31}\,\overline{x}_1\,\mathrm{e}^{u_1} + c_{32}\,\overline{x}_2\,\mathrm{e}^{u_2} + c_{33}\,\overline{x}_3\,\mathrm{e}^{u_3})}_{c_{31}\,\overline{x}_1\,(\mathrm{e}^{u_1}-1) + c_{32}\,\overline{x}_2\,(\mathrm{e}^{u_2}-1) + c_{33}\,\overline{x}_3\,(\mathrm{e}^{u_3}-1)} \\ &\quad = -\overline{x}_3\,(\mathrm{e}^{u_3} - 1)\Big(c_{31}\,(c_{13} + c_{33})\,\overline{x}_1\,(\mathrm{e}^{u_1} - 1) \\ &\qquad + c_{32}\,(c_{23} + c_{33})\,\overline{x}_2\,(\mathrm{e}^{u_2} - 1)\Big) - c_{33}^2\,\overline{x}_3^2\,(\mathrm{e}^{u_3} - 1)^2 \leq 0 \\ &\text{for all} \quad (u_1,u_2,u_3) \in \mathbb{R}^3, \end{aligned}$$

if

$$c_{13} = c_{23} = -c_{33}. \tag{1.37}$$

By Proposition 1.12 it follows that $(0,0,0)$ is a stable equilibrium state of the system (1.32) for $n = 3$, if the condition (1.37) is satisfied.

Further the system (1.24) for $n = 3$ has infinitely many solutions $(\overline{x}_1, \overline{x}_2, \overline{x}_3)$ with $\overline{x}_i > 0,\ i = 1,2,3$, which are stable equilibrium states of the system (1.26), (1.23), if the conditions (1.37) and

$$\begin{aligned} c_{23}\, c_1 + c_{13}\, c_2 &= 0, \\ c_{33}\, c_1 - c_{13}\, c_3 &< 0, \\ c_{33}\, c_2 - c_{23}\, c_3 &< 0 \end{aligned} \tag{1.38}$$

are satisfied. The first condition in (1.38) implies that

$$\overline{x}_3 = -\frac{c_1}{c_{13}} = -\frac{c_2}{c_{23}}.$$

From this it follows that the first two equations of (1.23) can be written in the form

$$\begin{aligned} \dot{x}_1(t) &= c_{13}\,(x_3(t) - \overline{x}_3)\; x_1(t), \\ \dot{x}_2(t) &= c_{23}\,(x_3(t) - \overline{x}_3)\; x_2(t). \end{aligned}$$

On using (1.37) we therefore conclude that

$$x_2(t) = \frac{x_2(0)}{x_1(0)}\, x_1(t) \tag{1.39}$$

and the system (1.23) can be replaced by (1.39) and

$$\begin{aligned} \dot{x}_1(t) &= (c_1 + c_{13}\, x_3(t))\; x_1(t), \\ \dot{x}_3(t) &= (c_3 + \widetilde{c}_{31}\, x_1(t) + c_{33}\, x_3(t))\; x_3(t). \end{aligned}$$

where

$$\widetilde{c}_{31} = c_{31} + c_{32}\, \frac{x_2(0)}{x_1(0)} < 0.$$

Finally, we get from

$$c_3 + \widetilde{c}_{31}\, \overline{x}_1 + c_{33}\, \overline{x}_3 = 0$$

that

$$\overline{x}_1 = -\frac{1}{\widetilde{c}_{31}} \left(c_3 - c_{33}\, \frac{c_1}{c_{13}} \right) > 0,$$

if

$$c_3 - c_{33}\, \frac{c_1}{c_{13}} > 0.$$

The last two cases are special cases of the following general situation:

Let $1 \leq m < n$ and let $X_1, \ldots, X_m$ be predator populations that have the populations $X_{m+1}, \ldots, X_n$ as prey and are neutral to each other.

Further we assume that $X_1, \ldots, X_m$ would decay exponentially in the absence of $X_{m+1}, \ldots, X_n$ and $X_{m+1}, \ldots, X_n$ would increase logistically in the absence of $X_1, \ldots, X_m$. Finally $X_{m+1}, \ldots, X_n$ are neutral to each other. All these assumptions lead to the following conditions:

$$\begin{aligned}
&c_1 < 0, \ \ldots \ , c_m < 0, \ c_{m+1} > 0, \ \ldots \ , c_n > 0, \\
&c_{11} = 0, \ \ldots \ , c_{mm} = 0, \ c_{m+1m+1} < 0, \ \ldots \ , c_{nn} < 0, \\
&c_{ij} = 0 \quad \text{for} \quad i, j = 1, \ldots, m, \ i \neq j. \\
&c_{ij} > 0 \quad \text{for} \quad i = 1, \ldots, m \quad \text{and} \quad j = m+1, \ldots, n, \\
&c_{ij} < 0 \quad \text{for} \quad i = m+1, \ldots, n \quad \text{and} \quad j = 1, \ldots, m, \\
&c_{ij} = 0 \quad \text{for} \quad i, j = m+1, \ldots, n, \ i \neq j.
\end{aligned}$$

The conditions (1.33) then read

$$\sum_{j=m+1}^{n} (-c_{ji}) > 0 \qquad \text{for} \quad i = 1, \ldots, m$$

and

$$\sum_{j=1}^{m} c_{ji} - c_{ii} > 0 \qquad \text{for} \quad i = m+1, \ldots, n$$

and are satisfied.

The conditions (1.35) turn out to be equivalent to

$$c_{ik} \sum_{j=m+1}^{n} (-c_{ji}) + c_{ki} \left(\sum_{j=1}^{m} c_{jk} - c_{kk} \right) = 0,$$
$$i = 1, \ldots, m, \ k = m+1, \ldots, n,$$

$$c_{ik} \left(\sum_{j=1}^{m} c_{ji} - c_{ii} \right) + c_{ki} \sum_{j=m+1}^{n} (-c_{jk}) = 0,$$
$$i = m+1, \ldots, n, \ k = 1, \ldots, m.$$

For $n = 3$ and $m = 1$ these conditions are equivalent to (1.36) and can be satisfied. For $n = 3$ and $m = 2$ they are equivalent to

$$c_{13} - c_{33} = 0 \qquad \text{and} \qquad c_{23} - c_{33} = 0$$

and are not satisfied.

We therefore define as Lyapunov function $V : \mathbb{R}^n \to \mathbb{R}$

$$V(u_1, \ldots, u_n) = \sum_{i=1}^{n} \sum_{j=m+1}^{n} (-c_{ji})\, \overline{x}_i \left(e^{u_i} - u_i \right) - \sum_{j=m+1}^{n} c_j$$
$$\text{for} \quad (u_1, \ldots, u_n) \in \mathbb{R}^n$$

and obtain

$$V(0,\dots,0)=0$$

and

$$V(u_1,\dots,u_n)>0 \qquad \text{for all} \quad (u_1,\dots,u_n)\neq(0,\dots,0)$$

because of

$$\sum_{j=m+1}^{n}(-c_{ji})>0 \qquad \text{for} \quad i=1,\dots,n.$$

The system (1.32) reads

$$\begin{aligned}
\dot{u}_i(t) &= \sum_{j=m+1}^{n} c_{ij}\,\overline{x}_j\left(\mathrm{e}^{u_j(t)}-1\right), \qquad i=1,\dots,m,\\
\dot{u}_i(t) &= \sum_{j=1}^{m} c_{ij}\,\overline{x}_j\left(\mathrm{e}^{u_j(t)}-1\right)+c_{ii}\,\overline{x}_i\left(\mathrm{e}^{u_i(t)}-1\right), \qquad i=m+1,\dots,n,
\end{aligned}$$

and we obtain

$$\begin{aligned}
\dot{V}(u_1,\dots,u_n) =& \sum_{i=1}^{m}\left(\sum_{k=m+1}^{n}(-c_{ki})\,\overline{x}_i\,(\mathrm{e}^{u_i}-1)\sum_{j=m+1}^{n}c_{ij}\,\overline{x}_j\,(\mathrm{e}^{u_j}-1)\right)\\
&+\sum_{i=m+1}^{n}\left(\sum_{k=m+1}^{n}(-c_{ki})\,\overline{x}_i\,(\mathrm{e}^{u_i}-1)\right)\left(\sum_{j=1}^{m}c_{ij}\,\overline{x}_j\,(\mathrm{e}^{u_j}-1)\right.\\
&\qquad\left.+c_{ii}\,\overline{x}_i\,(\mathrm{e}^{u_i}-1)\right)\\
=&\sum_{i=1}^{m}\sum_{j=m+1}^{n}\left(\sum_{k=m+1}^{n}(-c_{ki})\,c_{ij}\right.\\
&\qquad\left.+\sum_{k=m+1}^{n}(-c_{kj})\,c_{ji}\right)\overline{x}_i\overline{x}_j\,(\mathrm{e}^{u_i}-1)\,(\mathrm{e}^{u_j}-1)\\
&+\sum_{i=m+1}^{n}\sum_{k=m+1}^{n}(-c_{ki})\,c_{ii}\,\overline{x}_i^2\,(\mathrm{e}^{u_i}-1)^2\le 0,
\end{aligned}$$

if

$$\left(\sum_{k=m+1}^{n}c_{ki}\right)c_{ij}+\overbrace{\left(\sum_{k=m+1}^{n}c_{kj}\right)}^{c_{jj}}c_{ji}=0 \qquad \text{for} \quad i=1,\dots,m \quad \text{and} \quad j=m+1,\dots,n. \tag{1.40}$$

In the case $n = 3,\ m = 2$ these conditions are equivalent to (1.37). In the case $n = 3,\ m = 1$ they are equivalent to

$$(c_{21} + c_{31})\, c_{12} + c_{22}\, c_{21} = 0, \qquad (c_{21} + c_{31})\, c_{13} + c_{33}\, c_{31} = 0$$

and can be satisfied. The same holds true for the conditions (1.40).

1.2.5 Application to Evolution Matrix Games

In the following we give a mathematical definition of an evolution matrix game. As to a definition within a biological context we refer to [26] (see also Appendix A). We start with a bi-matrix game that is played with mixed strategies so that the strategy sets of the two players are given by two simplices

$$X = \left\{ x \in \mathbb{R}^m \mid x_i \geq 0 \ \text{ for } \ i = 1, \ldots, m \ \text{ and } \ \sum_{i=1}^{m} x_i = 1 \right\}$$

and

$$Y = \left\{ y \in \mathbb{R}^n \mid y_j \geq 0 \ \text{ for } \ j = 1, \ldots, n \ \text{ and } \ \sum_{j=1}^{n} y_j = 1 \right\}.$$

The corresponding payoff functions are given by

$$\Phi_1(x, y) = x^T A y \quad \text{and} \quad \Phi_2(x, y) = x^T B y \qquad \text{with} \quad x \in X \text{ and } y \in Y$$

for player 1 and 2, respectively, where A and B are $m \times n-$matrices. A pair $(\widehat{x}, \widehat{y}) \in X \times Y$ is called a Nash equilibrium, if

$$\widehat{x}^T A \widehat{y} \geq x^T A \widehat{y} \qquad \text{for all} \quad x \in X$$

and

$$\widehat{x}^T B \widehat{y} \geq \widehat{x}^T A y \qquad \text{for all} \quad y \in Y.$$

Now we assume that the game is symmetric, i.e.,

$$m = n \implies X = Y \qquad \text{and} \qquad B = A^T.$$

Then the two conditions for a Nash equilibrium are equivalent to

$$\widehat{x}^T A \widehat{y} \geq x^T A \widehat{y} \quad \text{and} \quad \widehat{y}^T A \widehat{x} \geq y^T A \widehat{x} \qquad \text{for all} \quad x, y \in X.$$

If $\widehat{x} = \widehat{y}$, these two conditions are equivalent to

$$\widehat{x}^T A \widehat{x} \geq x^T A \widehat{x} \qquad \text{for all} \quad x \in X.$$

In symmetric games the two players cannot be distinguished and only symmetric strategy pairs (x, x) are of interest. So there is only one player who plays against himself. But then the game has to be given another interpretation.

One interpretation could be that the player is represented by a population whose individuals have n strategies at their disposal in the struggle of life. Then the (mixed) strategies of the game can be considered as a probability distribution of the strategies of the individuals. These probability distributions are called population states and with this interpretation the game is called an evolution matrix game. The last condition for a Nash equilibrium of the form $(\widehat{x}, \widehat{x})$, $\widehat{x} \in X$ then leads to the following.

Definition. *A population state $\widehat{x} \in X$ is called a Nash equilibrium, if*

$$x^T A\widehat{x} \leq \widehat{x}^T A\widehat{x} \qquad \textit{for all} \quad x \in X.$$

In words that means that a deviation from $\widehat{x}$ does not lead to a higher payoff. For rational behavior this would suffice to maintain the population state $\widehat{x}$. However, animals do not behave rationally so that the stability of a Nash equilibrium is not guaranteed. This leads to the concept of an evolutionarily stable Nash equilibrium given by the

Definition. *A Nash equilibrium $\widehat{x} \in X$ is called evolutionarily stable, if $x^T A\widehat{x} = \widehat{x}^T A\widehat{x}$ for some $x \in X$ with $x \neq \widehat{x}$ implies that $x^T Ax < \widehat{x}^T Ax$.*

In words this means that, if a change from $\widehat{x}$ to x leads to the same payoff, x cannot be a Nash equilibrium.

Evolutionary stability can be characterized by a condition which is needed in a dynamical treatment of the game. In this connection we also need a necessary condition for a Nash equilibrium. in order to derive this we at first define for every $x \in X$ a support by

$$S(x) = \{i \in \{1, \ldots, n\} \mid x_i > 0\}.$$

Then we can prove the

Lemma 1.14. *If $\widehat{x} \in X$ is a Nash equilibrium, then it follows that*

$$e_i^T A\widehat{x} = \widehat{x}^T A\widehat{x} \qquad \textit{for all} \quad i \in S(\widehat{x})$$

where $e_i^T = (0, \ldots, 0, \underset{i}{1}, 0, \ldots, 0)$.

Proof. At first we have

$$\widehat{x}^T A\widehat{x} = \sum_{i=1}^{n} \widehat{x}_i e_i^T A\widehat{x} = \sum_{i \in S(\widehat{x})} \widehat{x}_i e_i^T A\widehat{x} \leq \max_{i \in S(\widehat{x})} e_i^T A\widehat{x}.$$

Since $\widehat{x}$ is a Nash equilibrium, it follows that

$$e_i^T A\widehat{x} \leq \widehat{x}^T A\widehat{x} \qquad \text{for all} \quad i \in S\left(\widehat{x}\right),$$

hence

$$\widehat{x}^T A\widehat{x} = \max_{i \in S(\widehat{x})} e_i^T A\widehat{x}.$$

Let

$$e_{i_0}^T A\widehat{x} = \max_{i \in S(\widehat{x})} e_i^T A\widehat{x}.$$

Then we obtain

$$0 = \widehat{x}^T A\widehat{x} - \sum_{i \in S(\widehat{x})} \widehat{x}_i e_i^T A\widehat{x} = \sum_{i \in S(\widehat{x})} \widehat{x}_i \left(e_{i_0}^T A\widehat{x} - e_i^T A\widehat{x}\right)$$

which implies

$$\widehat{x}^T A\widehat{x} = e_{i_0}^T A\widehat{x} = e_i^T A\widehat{x} \qquad \text{for all} \quad i \in S\left(\widehat{x}\right)$$

and concludes the proof.

The characterization of evolutionary stability is given by

Lemma 1.15. *A population state $\widehat{x} \in X$ is an evolutionarily stable Nash equilibrium, if and only if there exists some $\varepsilon > 0$ such that*

$$x^T Ax < \widehat{x}^T Ax \qquad \textit{for all} \quad x \in X \quad \textit{with} \quad x \neq \widehat{x} \quad \textit{and} \quad ||x - \widehat{x}||_2 < \varepsilon.$$

For the proof we refer to [22].

The game is stable by its very nature. For its application to problems of evolution, however, it is desirable to embed it into a dynamics in the course of time. In [22] we have chosen a time-discrete dynamical model that was first introduced in [26]. Here we choose a time-continuous dynamical model that has been introduced independently by Taylor and Jonker, Zuman and Hofbauer et. al. It is based on the following system of differential equations

$$\dot{x}_i(t) = x_i(t)\left((Ax(t))_i - x(t)^T Ax(t)\right), \quad t \in \mathbb{R}, \qquad (*)$$
$$\text{for} \quad i = 1, \ldots, n.$$

$x(t)$ denotes the state of the population at the time t, i.e., $x(t) \in X$. $x(t)^T Ax(t)$ can be considered as the average payoff in the population and $(Ax(t))_i$ the payoff to the i-th strategy. So the terms $(Ax(t))_i - x(t)^T Ax(t)$ can be interpreted as the growth rate of $x_i(t)$.

Now let $\widehat{x} \in X$ be a Nash equilibrium. Then by Lemma 1.14 $\widehat{x}$ is a solution of

$$\left(A\widehat{x}\right)_i - \widehat{x}^T A\widehat{x} = 0 \qquad \text{for all} \quad i \in \{1, \ldots, n\} \ \text{ with } \ \widehat{x}_i > 0.$$

This implies that $\widehat{x}$ is a rest point of the system $(*)$.

The question now arises under which condition this rest point is asymptotically stable. We shall show that this is the case, if $\widehat{x}$ is evolutionarily stable. For the proof we will make use of Proposition 1.12. So the first step consists of defining a Lyapunov function. For this purpose we define the set

$$X(\widehat{x}) = \{x \in X \mid S(x) \supseteq S(\widehat{x})\}$$

and the function

$$W(x) = \sum_{i \in S(\widehat{x})} \widehat{x}_i \ln x_i, \qquad x \in X(\widehat{x}).$$

Now let $x^* \in X(\widehat{x})$ be such that

$$W(x^*) \geq W(x) \qquad \text{for all} \quad x \in X(\widehat{x}).$$

Since $W = W(x)$, $x \in X(\widehat{x})$, is differentiable and strictly concave, this is equivalent with the existence of some $\lambda \in \mathbb{R}$, $\lambda \neq 0$, such that

$$\sum_{i \in S(\widehat{x})} \left(\frac{\widehat{x}_i}{x_i^*} + \lambda\right)(x_i^* - x_i) \geq 0 \qquad \text{for all} \quad x \in \mathbb{R}^n \text{ with } x_i \geq 0 \text{ for } i \in S(\widehat{x}).$$

This again is equivalent with

$$\frac{\widehat{x}_i}{x_i^*} + \lambda = 0 \iff \widehat{x}_i = -\lambda x_i^* \qquad \text{for all} \quad i \in S(\widehat{x})$$

which implies $\lambda = -1$ and $\widehat{x} = x^*$. Therefore, it follows that

$$W(\widehat{x}) \geq W(x) \qquad \text{for all} \quad x \in X(\widehat{x}).$$

If we define the function

$$V(x) = W(\widehat{x}) - W(x), \qquad x \in X(\widehat{x}),$$

then it follows that

$$V(x) \geq 0 \quad \text{for all} \quad x \in X(\widehat{x}) \qquad \text{and} \qquad V(x) = 0 \iff x = \widehat{x}.$$

Further it follows that

$$V_{x_i}(x) = -\frac{\widehat{x}_i}{x_i}, \quad i \in S(\widehat{x}), \qquad \text{for all} \quad x \in X(\widehat{x}).$$

If we put

$$f_i(x) = x_i\left((Ax)_i - x^T Ax\right), \qquad i \in S(\widehat{x}),$$

then it follows that

$$\nabla V(x)^T f(x) = -\widehat{x}^T Ax + x^T Ax \qquad \text{for all} \quad x \in X(\widehat{x}).$$

If we now assume that $\widehat{x}$ is evolutionarily stable, then it follows by Lemma 1.15 that there exists some $\varepsilon > 0$ such that

$$\nabla V(x)^T f(x) < 0 \qquad \text{for all} \quad x \in X(\widehat{x}) \text{ with } x \neq \widehat{x} \text{ and } \|x - \widehat{x}\|_2 < \varepsilon.$$

Hence Proposition 1.12 with $X = \{x \in X(\widehat{x}) \mid \|x - \widehat{x}\|_2 < \varepsilon\}$ implies that $\widehat{x}$ is asymptotically stable.

1.3 Time-Discrete Dynamical Systems

1.3.1 The Autonomous Case: Definitions and Elementary Properties

Let X be a metric space with metric $d : X \times X \to \mathbb{R}_+$ and let $f : X \to X$ be a continuous mapping. If we define a mapping $\pi : X \times \mathbb{N}_0 \to X$, by

$$\pi(x,k) = f^k(x) = \underbrace{f \circ \ldots \circ f}_{k-\text{times}}(x), \qquad x \in X, \quad k \in \mathbb{N}_0, \tag{1.41}$$

then π is continuous and it follows

$$\pi(x,0) = x \qquad \text{for all} \quad x \in X$$

and

$$\pi(\pi(x,k),l) = f^l\left(f^k(x)\right) = f^{l+k}(x) = \pi(x,k+l) \qquad \text{for all} \quad x \in X.$$

Hence by the definition (1.41) we obtain a time-discrete dynamical system.

Conversely, if $\pi : X \times \mathbb{N}_0 \to X$ is a time-discrete flow and if we define a continuous mapping $f : X \to X$ by

$$f(x) = \pi(x,1) \qquad \text{for all} \quad x \in X,$$

then it follows that

$$f^k(x) = \pi(x,k) \qquad \text{for all} \quad x \in X \quad \text{and} \quad k \in \mathbb{N}_0.$$

Without loss of generality we therefore can consider a time-discrete dynamical system as a pair (X,f) where X is a metric space and $f : X \to X$ is a continuous mapping. The flow is then defined by (1.41) and the system is called autonomous in contrast to non-autonomous systems which will be defined in Section 1.3.8.

For every $x \in X$ we define an orbit starting with x by

$$\gamma_f(x) = \bigcup_{n \in \mathbb{N}_0} \{f^n(x)\} \tag{1.42}$$

(see (1.1)). Further we define as limit set of $\gamma_f(x)$ the set

$$L_f(x) = \bigcap_{n \in \mathbb{N}_0} \overline{\bigcup_{m \geq n} \{f^m(x)\}} \tag{1.43}$$

(see (1.2)). This limit set can be given by an equivalent definition which is the content of

Proposition 1.16. *For every $x \in X$ the limit set $L_f(x)$ being defined by (1.43) consists of all accumulation points of the sequence $(f^n(x))_{n\in\mathbb{N}_0}$.*

Proof.
1) Let $y \in X$ be an accumulation point of $(f^n(x))_{n\in\mathbb{N}_0}$. Then there exists a subsequence $(f^{n_i}(x))_{i\in\mathbb{N}_0}$ with $f^{n_i}(x) \to y$. This implies that, for every $n \in \mathbb{N}_0$,

$$y \in \overline{\bigcup_{m\geq n} \{f^m(x)\}}$$

which in turn implies that $y \in L_f(x)$.
2) Let $y \in L_f(x)$. Then

$$y \in \overline{\bigcup_{m\geq n} \{f^m(x)\}} \qquad \text{for every} \quad n \in \mathbb{N}_0.$$

Therefore, for every $n \in \mathbb{N}_0$, there is a sequence $\left(f^{k_i+n}(x)\right)_{i\in\mathbb{N}_0}$ with $f^{k_i+n}(x) \to y$ as $i \to \infty$. Hence, for every $n \in \mathbb{N}_0$, there exists an $i_n \in \mathbb{N}_0$ such that

$$d\left(f^{k_i+n}(x), y\right) \leq \frac{1}{n} \qquad \text{for all} \quad i \geq i_n$$

and we can assume that $i_{n+1} > i_n$. This implies that $\left(f^{k_{i_n}+n}(x)\right)_{n\in\mathbb{N}}$ is a subsequence of $(f^n(x))_{n\in\mathbb{N}_0}$ with $f^{k_{i_n}+n}(x) \to y$ as $n \to \infty$. This means that y is an accumulation point of $(f^n(x))_{n\in\mathbb{N}_0}$ which completes the proof. □

Definition. *A non-empty subset $H \subseteq X$ is called positively (negatively) invariant (with respect to a mapping $f : X \to X$), if*

$$f(H) \subseteq H \qquad (H \subseteq f(H)),$$

and invariant, if

$$f(H) = H.$$

Exercise 1.1. Show that for a continuous mapping $f : X \to X$ the following holds true:

(a) The closure of a positively invariant subset of X is also positively invariant (with respect to f).
(b) The closure of a relatively compact invariant subset of X is also invariant (with respect to f).

According to Proposition 1.16 the limit set $L_f(x)$ for some $x \in X$ given by (1.43) can be empty, if the sequence $(f^n(x))_{n\in\mathbb{N}_0}$ does not have accumulation points. If this is not the case, then we have the

Proposition 1.17. *If, for some $x \in X$, the limit set $L_f(x)$ given by (1.43) is non-empty, then it is closed and positively invariant.*

Proof. The closedness of $L_f(x)$ is an immediate consequence of the definition (1.43). Let $y \in L_f(x)$ be given. Then, by Proposition 1.16, there exists a subsequence $(f^{n_i}(x))_{i\in\mathbb{N}_0}$ of the sequence $(f^n(x))_{n\in\mathbb{N}_0}$ with $f^{n_i}(x) \to y$ as $i \to \infty$. This implies, due to the continuity of f, that $f^{n_i+1}(x) \to f(y)$, hence $f(y) \in L_f(x)$. This shows $f(L_f(x)) \subseteq L_f(x)$, i.e. that $L_f(x)$ is positively invariant. □

If X is compact, then, for every $x \in X$, the limit set $L_f(x)$ given by (1.43) is non-empty and we can even prove

Proposition 1.18. *If X is compact, then, for every $x \in X$, the limit set $L_f(x)$ given by (1.43) is compact, invariant and the smallest closed subset $S \subseteq X$ with*

$$\lim_{n\to\infty} \varrho(f^n(x), S) = 0 \qquad \textit{where} \quad \varrho(y, S) = \min\{d(y,z) \mid z \in S\}. \tag{1.44}$$

Proof. As a non-empty closed subset of the compact metric space X the limit set $L_f(x)$ is also compact.

In order to show the invariance of $L_f(x)$ it suffices to show that $L_f(x) \subseteq f(L_f(x))$. Let $y \in L_f(x)$ be given. Then there exists a subsequence $(f^{n_i}(x))_{i\in\mathbb{N}_0}$ of the sequence $(f^n(x))_{n\in\mathbb{N}_0}$ with $f^{n_i}(x) \to y$ and with no loss generality we can assume that there is some $z \in L_f(x)$ with $f^{n_i-1}(x) \to z$. By the continuity of f this implies $f^{n_i}(x) \to f(z)$, hence $y = f(z) \in L_f(x)$ and therefore $L_f(x) \subseteq f(L_f(x))$. In order o show (1.44) we at first show that

$$\lim_{n\to\infty} \varrho(f^n(x), L_f(x)) = 0.$$

For that purpose we assume that

$$\varrho(f^n(x), L_f(x)) \not\to 0 \qquad \text{as} \quad n \to \infty.$$

Then there is a subsequence $(f^{n_i}(x))_{i\in\mathbb{N}_0}$ of $(f^n(x))_{n\in\mathbb{N}_0}$ with

$$\varrho(f^{n_i}(x), L_f(x)) \not\to 0 \qquad \text{as} \quad i \to \infty,$$

and some $y \in L_f(x)$ with $\lim_{i\to\infty} d(f^{n_i}(x), y) = 0$ which implies

$$\lim_{i\to\infty} \varrho(f^{n_i}(x), L_f(x)) = 0$$

and leads to a contradiction.

Now let $S \subseteq X$ be any closed subset with

$$\lim_{n\to\infty} \varrho(f^n(x), S) = 0.$$

Then we choose any $y \in L_f(x)$ and conclude by Proposition 1.16 the existence of a subsequence $(f^{n_i}(x))_{i\in\mathbb{N}_0}$ of $(f^n(x))_{n\in\mathbb{N}_0}$ with $f^{n_i}(x) \to y$.

Further it follows that

$$\lim_{i \to \infty} \varrho\left(f^{n_i}(x), S\right) = 0,$$

which implies $y \in S$, since S is closed. This completes the proof of Proposition 1.18. □

Definition. *A closed invariant subset of X is called invariantly connected, if it is not representable as a disjoint union of two non-empty, invariant and closed subsets of X.*

Definition. *A sequence $(f^n(x))_{n \in \mathbb{N}_0}$, $x \in X$, is called periodic or cyclic, if there exists a number $k \in \mathbb{N}$ with $f^k(x) = x$. The smallest number with this property is called the period of the sequence. If $k = 1$, then $x \in X$ is called a fixed point of $f : X \to X$.*

Exercise 1.2. Show that a finite subset $H \subseteq X$ is invariantly connected, if and only if for every $x \in H$ the sequence $(f^n(x))_{n \in \mathbb{N}_0}$ is periodic and its period is equal to the cardinality of H.

Proposition 1.18 can be supplemented by

Proposition 1.19. *If X is compact, then, for every $x \in X$, the limit set $L_f(x)$ given by (1.43) is invariantly connected.*

Proof. Let us assume that, for some $x \in X$, there exist two non-empty, invariant and closed subsets A_1 and A_2 of $L_f(x)$ which are disjoint and satisfy $A_1 \cup A_2 = L_f(x)$.

Since $L_f(x)$ is compact, A_1 and A_2 are also compact and there exist disjoint open subsets U_1 and U_2 of X with $A_1 \subseteq U_1$ and $A_2 \subseteq U_2$. Since f is uniformly continuous on A_1, there is an open subset V_1 of X with $A_1 \subseteq V_1$ and $f(V_1) \subseteq U_1$.

Since $L_f(x)$ is the smallest closed set $S \subseteq X$ with (1.44), the sequence $(f^n(x))_{n \in \mathbb{N}_0}$ must intersect V_1 as well as U_2 infinitely many times. This implies the existence of a subsequence $(f^{n_i}(x))_{i \in \mathbb{N}_0}$ which is neither contained in V_1 nor U_2 and which can be assumed to be convergent. This, however, is impossible and leads to a contradiction to the assumption that $L_f(x)$ is not invariantly connected. □

Exercise 1.3. Given a compact and positively invariant subset $K \subseteq X$. Show that $\bigcap_{n \in \mathbb{N}_0} f^n(K)$ with $f^n(K) = \{f^n(x) \mid x \in K\}$ for all $n \in \mathbb{N}_0$ is non-empty, compact and the largest invariant subset of K.

1.3.2 Localization of Limit Sets with the Aid of Lyapunov Functions

Let X be a metric space and let $f : X \to X$ be a continuous mapping. Further let $G \subseteq X$ be a non-empty subset.

Definition. *A function $V : X \to \mathbb{R}$ is called a Lyapunov function with respect to f on G, if*

(1) *V is continuous on X,*
(2) *$V(f(x)) - V(x) \leq 0$ for all $x \in G$.*

If $V : X \to \mathbb{R}$ is a Lyapunov function with respect to f on G, then we define the set

$$E = \left\{x \in X \mid V(f(x)) = V(x),\ x \in \overline{G}\right\}$$

where $\overline{G}$ is the closure of G. Further we put, for every $c \in \mathbb{R}$,

$$V^{-1}(c) = \{x \in X \mid V(x) = c\}.$$

Then we prove

Proposition 1.20. *Let $G \subseteq X$ be non-empty and relatively compact. Further let V be a Lyapunov function with respect to f on G and finally let $x_0 \in G$ be such that*

$$f^n(x_0) \in G \qquad \text{for all} \quad n \in \mathbb{N}.$$

Then there exists some $c \in \mathbb{R}$ such that

$$L_f(x_0) \subseteq M \cap V^{-1}(c)$$

where M is the largest invariant subset of E.

Proof. If we define $x_n = f^n(x_0)$, $n \in \mathbb{N}_0$, then the sequence $\left(V(x_n)_{n \in \mathbb{N}_0}\right)$ is contained in $V\left(\overline{G}\right)$ and therefore bounded.

Further it follows that

$$V(x_{n+1}) \leq V(x_n) \qquad \text{for all} \quad n \in \mathbb{N}_0.$$

Therefore there exists some $c \in \mathbb{R}$ with $c = \lim\limits_{n\to\infty} V(x_n)$.

Now let $p \in L_f(x_0)$. Then there exists a subsequence $(x_{n_i})_{i \in \mathbb{N}_0}$ of $(x_n)_{n \in \mathbb{N}_0}$ with $x_{n_i} \to p$ which implies $c = \lim\limits_{i\to\infty} V(x_{n_i}) = V(p)$ due to the continuity of V. Hence $p \in V^{-1}(c)$ and thus $L_f(x_0) \subseteq V^{-1}(c)$. Since, by Proposition 1.18, $L_f(x_0)$ is invariant, it follows that $V(f(p)) = c$ for all $p \in L_f(x_0)$ and hence $V(f(p)) = V(p)$ for all $p \in L_f(x_0)$ which implies $L_f(x_0) \subseteq E$ and in turn $L_f(x_0) \subseteq M$. This completes the proof. □

Let us demonstrate this result by an example:

We take $X = \mathbb{R}^2$ equipped with the Euclidian norm and define $f : X \to X$ by

$$f(x,y) = (f_1(x,y), f_2(x,y))^T, \qquad (x,y) \in X,$$

with

$$f_1(x,y) = \frac{y}{1+x^2}, \qquad f_2(x,y) = \frac{x}{1+y^2}. \tag{1.45}$$

Further we choose

$$V(x,y) = x^2 + y^2, \qquad (x,y) \in X. \tag{1.46}$$

Then it follows that

$$V(f(x,y)) - V(x,y) = \left(\frac{1}{(1+y^2)^2} - 1\right) x^2 + \left(\frac{1}{(1+x^2)^2} - 1\right) y^2 \leq 0,$$
$$\text{for all} \quad (x,y) \in X.$$

This shows that V is a Lyapunov function with respect to f on $X = \mathbb{R}^2$. Further we see that

$$E = \{(x,0) \mid x \in \mathbb{R}\} \cup \{(0,y) \mid y \in \mathbb{R}\}.$$

From $f(x,0) = (0,x)$ for all $x \in \mathbb{R}$ and $f(0,y) = (y,0)$ for all $y \in \mathbb{R}$, it follows that E is invariant and hence $M = E$.

Further we conclude

$$f^2(x,0) = f(0,x) = (x,0) \qquad \text{for all} \quad x \in \mathbb{R}$$

and

$$f^2(0,y) = f(y,0) = (0,y) \qquad \text{for all} \quad y \in \mathbb{R}.$$

Now let $G = \left\{(x,y) \in \mathbb{R}^2 \mid x^2 + y^2 < r\right\}$ for any $r > 0$.

Then it follows that $f(G) \subseteq G$ and for every $(x_0, y_0) \in G$ there exists some $c \in \mathbb{R}$ such that

$$L_f(x_0, y_0) \subseteq E \cap \left\{(x,y) \mid x^2 + y^2 = c^2\right\} = \{(c,0), (0,c)\}.$$

Since $L_f(x_0, y_0)$ is invariantly connected by Proposition 1.19 it follows from Exercise 1.2 that

$$L_f(x_0, y_0) = \{(c,0), (0,c)\}.$$

1.3.3 Stability Based on Lyapunov's Method

Let $f : X \to X$ be a continuous mapping where X is a metric space.

Definition. *A relatively compact set $H \subseteq X$ is called stable with respect to f, if for every relatively compact open set $U \subseteq X$ with $U \supseteq \overline{H}$ = closure of H there exists an open set $W \subseteq X$ with $\overline{H} \subseteq W \subseteq U$ such that*

$$f^n(W) \subseteq U \qquad \text{for all} \quad n \in \mathbb{N}_0$$

where

$$f^n(W) = \{f^n(x) \mid x \in W\}.$$

Theorem 1.1. *Let $H \subseteq X$ be relatively compact and such that for every relatively compact open set $U \subseteq X$ with $U \supseteq \overline{H}$ there exists an open subset B_U of U with $B_U \supseteq \overline{H}$ and $f(B_U) \subseteq U$.*

Further let $G \subseteq X$ be an open set with $G \supseteq \overline{H}$ such that there exists a Lyapunov function V with respect to f on G which is positive definite with respect to $\overline{H}$, i.e.,

$$V(x) \geq 0 \quad \text{for all} \quad x \in G \qquad \text{and} \qquad \left(V(x) = 0 \iff x \in \overline{H}\right).$$

Then H is stable with respect to f.

Proof. Let $U \subseteq X$ be an arbitrary relatively compact open set with $U \supseteq \overline{H}$. Then $U^* = U \cap G$ is also a relatively compact open set with $U^* \supseteq \overline{H}$ and there exists an open set $B_{U^*} \subseteq U^*$ with $B_{U^*} \supseteq \overline{H}$ and $f(B_{U^*}) \subseteq U^*$.

Let us put

$$m = \min\left\{V(x) \mid x \in \overline{U^*} \setminus B_{U^*}\right\}.$$

Since $\overline{H} \cap \left(\overline{U^*} \setminus B_{U^*}\right)$ is empty, it follows that $m > 0$. If we define

$$W = \{x \in U^* \mid V(x) < m\},$$

then W is open and $\overline{H} \subseteq W \subseteq B_{U^*}$.

Now let $x \in W$ be chosen arbitrarily. Then $x \in B_{U^*}$ and therefore $f(x) \in U^*$. Further we have

$$V(f(x)) \leq V(x) < m,$$

hence $f(x) \in W \subseteq B_{U^*}$. This implies $f^2(x) = f(f(x)) \in U^*$ and

$$V\left(f^2(x)\right) \leq V(f(x)) < m, \qquad \text{hence} \quad f^2(x) \in W.$$

By induction it therefore follows that

$$f^n(x) \in W \subseteq U^* \subseteq U \qquad \text{for all} \quad n \in \mathbb{N}_0.$$

this shows that H is stable with respect to f. □

Exercise 1.4.
(a) Give an explicit definition of the instability of a relatively compact set $H \subseteq X$ with respect to f as logical negation of the notion of stability.

(b) Show: If a relatively compact set $H \subseteq X$ is stable with respect to f, then its closure $\overline{H}$ is positively invariant, i.e. $f\left(\overline{H}\right) \subseteq \overline{H}$.

Definition. *A set $H \subset X$ is called an attractor with respect to f, if there exists an open set $U \subseteq X$ with $U \supseteq \overline{H}$ such that*

$$\lim_{n\to\infty} \varrho\left(f^n(x), \overline{H}\right) = 0 \quad \left(\textit{in short: } f^n(x) \to \overline{H}\right) \qquad \textit{for all} \quad x \in U$$

where

$$\varrho\left(y, \overline{H}\right) = \inf\left\{d(y,z) \mid z \in \overline{H}\right\}.$$

If $H \subseteq X$ is stable and an attractor with respect to f, then H is called asymptotically stable with respect to f.

Theorem 1.2. *Let $H \subseteq X$ be such that there exists a relatively compact open set $U \supseteq \overline{H}$ with $f(U) \subseteq U$.*

Further let $V : X \to \mathbb{R}$ be a Lyapunov function with respect to f on U which is positive definite with respect to $\overline{H}$, i.e.

$$V(x) \geq 0 \quad \textit{for all} \quad x \in U \qquad \textit{and} \qquad \left(V(x) = 0 \iff x \in \overline{H}\right).$$

Finally let

$$\lim_{n\to\infty} V\left(f^n(x)\right) = 0 \qquad \textit{for all} \quad x \in U.$$

Then H is an attractor with respect to f.

Proof. Let $x \in U$ be chosen arbitrarily. Since the sequence $\left(f^n(x)\right)_{n\in\mathbb{N}_0}$ is contained in U, for every subsequence $\left(f^{n_i}(x)\right)_{i\in\mathbb{N}_0}$ of $\left(f^n(x)\right)_{n\in\mathbb{N}_0}$ there exists a subsequence $\left(f^{n_{i_j}}(x)\right)_{j\in\mathbb{N}_0}$ and some $q \in U$ with

$$\lim_{j\to\infty} f^{n_{i_j}}(x) = q.$$

This implies

$$\lim_{j\to\infty} V\left(f^{n_{i_j}}(x)\right) = V(q) = 0,$$

hence $q \in \overline{H}$, and therefore $f^{n_{i_j}}(x) \to \overline{H}$ as $j \to \infty$. From this it follows that $f^n(x) \to \overline{H}$ which shows that H is an attractor with respect to f. □

Corollary. *Under the assumptions of Theorem 1.1 and 1.2 it follows that $H \subseteq X$ is asymptotically stable.*

As an important special case we prove the following

Theorem 1.3. *Let $x^* \in X$ be a fixed point of f, i.e. $f(x^*) = x^*$. Further let there exist an open set $G \subseteq X$ with $x^* \in G$ and a Lyapunov function V with respect to f on G which is positive definite with respect to x^*, i.e.*

$$V(x) \geq 0 \quad \text{for all} \quad x \in G \qquad \text{and} \qquad (V(x) = 0 \iff x = x^*).$$

Then $\{x^*\}$ is stable with respect to f.

If in addition

$$V(f(x)) < V(x) \qquad \text{for all} \quad x \in G \quad \text{with} \quad x \neq x^*, \tag{1.47}$$

then $\{x^*\}$ is asymptotically stable with respect to f.

Proof. Let $U \subseteq X$ be a relatively compact open set with $x^* \in U$. Then there exists some $r > 0$ such that

$$B_r(x^*) = \{x \in X \mid d(x, x^*) < r\} \subseteq U.$$

Since f is continuous in x^*, there exists some $s \in (0, r)$ such that $f(B_s(x^*)) \subseteq B_r(x^*)$. Hence, if we put $B_U = B_s(x^*)$, then $f(B_U) \subseteq U$, B_U is open and $\{x^*\} \subseteq B_U$. By Theorem 1.1 therefore $\{x^*\}$ is stable with respect to f.

From the stability part of the proof of Theorem 1.1 we deduce for every relatively compact open subset U of X the existence of an open subset W of $U \cap G$ with $x^* \in W$, $\overline{W} \subseteq G$ and

$$f^n(x) \in W \qquad \text{for all} \quad x \in W \quad \text{and all} \quad n \in \mathbb{N}_0.$$

Now let $x \in W$ be chosen arbitrarily.

Assumption: $f^n(x) \not\to x^*$.
Then there exists a subsequence $(f^{n_k}(x))_{k \in \mathbb{N}_0}$ and some $\overline{x} \in \overline{W}$ with $\overline{x} \neq x^*$ and

$$\lim_{k \to \infty} f^{n_k}(x) = \overline{x}. \tag{$*$}$$

$f^{n_k}(x) \in W$ for all $k \in \mathbb{N}_0$ implies $f(f^{n_k}(x)) \in W$ for all $k \in \mathbb{N}_0$ and hence

$$f(f^{n_k}(x)) \to f(\overline{x}) \in \overline{W} \subseteq G.$$

This implies the existence of a neighborhood $B = \{z \in W \mid d(z, \overline{x}) \leq \varepsilon\}$ of $\overline{x}$ with $x^* \notin B$ and $f(B) \subseteq G$. For every $z \in B$ we therefore have $V(f(z)) < V(z)$ which implies

$$q = \sup_{z \in B} \frac{V(f(z))}{V(z)} < 1, \qquad \text{hence} \quad V(f(z)) \leq q\, V(z) \quad \text{for all} \quad z \in B.$$

Because of $(*)$ there is some $k_0 \in \mathbb{N}_0$ with $f^{n_k}(x) \in B$ for all $k \geq k_0$.

This implies, for all $k \geq k_0$, because of $f^n(f^{n_k}(x)) \in W \subseteq B$ for all $n \in \mathbb{N}$ that

$$V\left(f^{n_{k+1}}(x)\right) = V\left(f^{n_{k+1}-n_k-1}\left(f^{n_k+1}(x)\right)\right) \leq V\left(f^{n_k+1}(x)\right) \leq q\, V\left(f^{n_k}(x)\right)$$

and by iteration it follows that

$$V\left(f^{n_{k_0}+l}(x)\right) \leq q^l\, V\left(f^{n_{k_0}}(x)\right) \qquad \text{for all} \quad l \in \mathbb{N}.$$

This implies

$$\lim_{l\to\infty} V\left(f^{n_{k_0}+l}(x)\right) = 0 \qquad \text{and therefore} \quad V(\overline{x}) = 0$$

which is impossible because of $\overline{x} \neq x^*$. Thus the above assumption is false and hence

$$f^n(x) \to x^* \qquad \text{for all} \quad x \in W$$

which shows that x^* is an attractor with respect to f. □

Let us demonstrate this result by the same example as in Section 1.3.2, i.e. we choose $f : \mathbb{R}^2 \to \mathbb{R}^2$ as

$$f(x,y) = (f_1(x,y), f_2(x,y)), \qquad (x,y) \in X = \mathbb{R}^2,$$

with $f_1 = f_1(x,y)$ and $f_2 = f_2(x,y)$ defined by (1.45).

If we choose $V : X \to \mathbb{R}$ in the form of (1.46), then V is a Lyapunov function with respect to f on $G = X = \mathbb{R}^2$.

If we choose $x^* = (0,0)^T$, then V is positive definite with respect to x^* and x^* is a fixed point of f. Further (1.47) is satisfied.

By Theorem 1.3 it follows that $x^* = (0,0)^T$ is asymptotically stable with respect to f.

With the aid of Proposition 1.20 we can prove

Theorem 1.4. *Let $G \subseteq X$ be open, relatively compact and positively invariant with respect to f, i.e. $f(G) \subseteq G$. Further let V be a Lyapunov function with respect to f on G. Then the largest invariant subset M of*

$$E = \left\{x \in \overline{G} \mid V(f(x)) = V(x)\right\}$$

is an attractor with respect to f.

If in addition V is constant on M, then M is asymptotically stable with respect to f.

Proof. Let us put $U = G$. If we choose $x_0 \in U$ arbitrarily, then the sequence $\left(f^n(x_0)\right)_{n\in\mathbb{N}_0}$ is contained in G, since G is positively invariant. Proposition 1.20 implies $L_f(x_0) \subseteq M$. We have to show that $\lim\limits_{n\to\infty} \varrho\left(f^n(x_0), \overline{M}\right) = 0$.

Now let $(f^{n_i}(x_0))_{i\in\mathbb{N}_0}$ be an arbitrary subsequence of $(f^n(x_0))_{n\in\mathbb{N}_0}$. Then there is a subsequence $(f^{n_{i_j}}(x_0))_{j\in\mathbb{N}_0}$ and an element $y \in L_f(x_0) \subseteq M$ with $f^{n_{i_j}}(x_0) \to y$ which implies that $\lim\limits_{j\to\infty} \varrho\left(f^{n_{i_j}}(x_0), \overline{M}\right) = 0$ and in turn $\varrho\left(f^n(x_0), \overline{M}\right) \to 0$.

In order to show that M is asymptotically stable with respect to f we have to show that M is stable with respect to f. For that purpose we apply Theorem 1.1 and verify its assumptions. At first let $U \subseteq X$ be relatively compact, open and such that $U \supseteq \overline{H}$. If we define

$$B_U = \{x \in U \cap G \mid f(x) \in U\},$$

then B_U is open, $B_U \supseteq \overline{M}$, $B_U \subseteq U$, and $f(B_U) \subseteq U$.

Let

$$V(x) = c \qquad \text{for all} \quad x \in M. \tag{1.48}$$

Let us assume that there is some $x \in G$ with $V(x) < c$.

Then it follows that

$$V(f^n(x)) \le V(x) < c \qquad \text{for all} \quad n \in \mathbb{N}.$$

Since $L_f(x) \subseteq M$, there exists some $y \in L_f(x) \subseteq M$ with $V(y) < c$ which contradicts (1.48). Hence

$$V(x) \ge c \qquad \text{for all} \quad x \in G$$

and further

$$V(x) = c \quad \Longleftrightarrow \quad x \in \overline{M}.$$

Therefore, if we define

$$\widetilde{V}(x) = V(x) - c, \qquad x \in X,$$

then we obtain a Lyapunov function with respect to f on G which is positive definite with respect to $\overline{M}$. The assertion now follows by Theorem 1.1. □

Let us again demonstrate this result by the above example (1.45) with Lyapunov function (1.46). Let again

$$G = \left\{(x,y) \in \mathbb{R}^2 \mid x^2 + y^2 < r^2\right\} \qquad \text{for some} \quad r > 0.$$

Then G is open, relatively compact and positively invariant with respect to f. From Theorem 1.4 it therefore follows that

$$M = \left(\{(x,0) \mid x \in \mathbb{R}\} \cup \{(0,x) \mid x \in \mathbb{R}\}\right) \cap G$$

is an attractor with respect to f. It even follows that for every $x_0 \in G$ it is true that

$$\lim_{n\to\infty} \varrho\left(f^n(x_0), \overline{M}\right) = 0.$$

1.3.4 Stability of Fixed Points via Linearisation

Let us assume that X is a non-empty open subset of a normed linear space $(E, ||\cdot||)$ and $f : X \to X$ is a continuous mapping which is continuously Fréchet differentiable at every $x \in X$. We denote the Fréchet derivative of f at $x \in X$ by f'_x which is a continuous linear mapping $f'_x : E \to E$ whose norm is given by

$$||f'_x|| = \sup\{||f'_x(h)|| \mid h \in E \text{ with } ||h|| = 1\}$$

for $x \in X$. Then we can prove

Theorem 1.5. *Let $x^* \in X$ be a fixed point of f, i.e. $f(x^*) = x^*$. Then the following two statements are true:*

a) Let $||f'_{x^*}|| < 1$. Then there is some $\varepsilon > 0$ and some $c \in [0,1)$ such that

$$x_0 \in X \text{ and } ||x_0 - x^*|| < \varepsilon \implies ||f^n(x_0) - x^*|| \leq c^n ||x_0 - x^*|| \quad \text{for all } n \in \mathbb{N}$$

which implies that x^* is asymptotically stable with respect to f (Exercise).

b) Let f'_{x^*} be continuously invertible and $||{f'_{x^*}}^{-1}||^{-1} > 1$. Then there is some $\delta > 0$ and some $d > 1$ such that

$$f^k(x_0) \in X \quad \text{and} \quad ||f^k(x_0) - x^*|| < \delta \qquad \text{for} \quad 0 \leq k \leq n-1$$

implies that

$$||f^n(x_0) - x^*|| \geq d^n ||x_0 - x^*||$$

for all $n \in \mathbb{N}$. This implies that x^* is unstable with respect to f (Exercise, see Exercise 1.4 a)).

Proof. a) From the continuity of $x \to ||f'_x||$, $x \in X$, it follows that there exists some $\varepsilon > 0$ and some $c \in [0,1)$ such that

$$||f'_x|| \leq c \qquad \text{for all} \quad x \in B(x^*, \varepsilon) = \{y \in E \mid ||y - x^*|| < \varepsilon\}.$$

The mean value theorem implies

$$||f(x) - f(y)|| \leq ||f'_z|| \, ||x - y||$$

for all $x, y \in B(x^*, \varepsilon)$ and some $z = \alpha x + (1-\alpha) y$ with $\alpha \in (0,1)$.

This implies

$$||f(x) - f(y)|| \leq c||x - y|| \qquad \text{for all} \quad x, y \in B(x^*, \varepsilon)$$

and in particular

$$||f(x) - x^*|| \leq c||x - x^*|| < \varepsilon \qquad \text{for all} \quad x \in B(x^*, \varepsilon).$$

Therefore it follows that $f(B(x^*,\varepsilon)) \subseteq B(x^*,\varepsilon)$ and hence

$$\begin{aligned}||f^n(x) - x^*|| = ||f \circ f^{n-1}(x_0) - f(x^*)|| &\leq c||f^{n-1}(x_0) - x^*|| \\ &\leq \ldots \leq c^n ||x_0 - x^*|| \\ &\text{for all } \ x_0 \in B(x^*,\varepsilon) \ \text{ and } \ n \in \mathbb{N}.\end{aligned}$$

b) For every $x, y \in X$ we have

$$||x-y|| = ||{f'_{x^*}}^{-1}(f'_{x^*}(x)) - {f'_{x^*}}^{-1}(f'_{x^*}(y))|| \leq ||{f'_{x^*}}^{-1}|| \, ||f'_{x^*}(x-y)||$$

which implies

$$||f'_{x^*}(x-y)|| \geq d' \, ||x-y|| \qquad \text{with} \quad d' = ||{f'_{x^*}}^{-1}||^{-1} > 1.$$

From the Fréchet differentiability it follows that

$$f(x) - f(x^*) = f'_{x^*}(x - x^*) + \varepsilon(||x - x^*||) \qquad \text{for} \quad x \in X$$

where

$$\lim_{x \to x^*} \frac{||\varepsilon(||x - x^*||)||}{||x - x^*||} = 0.$$

If one chooses $\eta > 0$ with $d = d' - \eta > 1$ and $\delta > 0$ such that

$$\frac{||\varepsilon(||x - x^*||)||}{||x - x^*||} \leq \eta \qquad \text{for all} \quad B(x^*,\delta) \setminus \{x^*\},$$

then it follows that

$$\begin{aligned}||f(x) - f(x^*)|| &\geq ||f'_{x^*}(x-y)|| - ||\varepsilon(||x - x^*||)|| \\ &\geq d' \, ||x - x^*|| - \eta \, ||x - x^*|| \\ &= d \, ||x - x^*||\end{aligned}$$

for all $x \in B(x^*,\delta)$. Now let $x_0 \in B(x^*,\delta)$ be such that

$$||f^k(x_0) - x^*|| < \delta \qquad \text{for} \quad 0 \leq k \leq n-1, \ n \in \mathbb{N}.$$

Then it follows that

$$\begin{aligned}||f^n(x_0) - x^*|| = ||f \circ f^{n-1}(x_0) - f(x^*)|| &\geq d \, ||f^{n-1}(x_0) - x^*|| \\ &\geq \ldots \geq d^n \, ||x_0 - x^*||.\end{aligned}$$

This completes the proof of Theorem 1.5. □

Let us consider the following important special case:
We assume $E = \mathbb{R}^k$ equipped with any norm $||\cdot||$. Let again $X \subseteq E$ be non-empty and open and let again $f : X \to X$ be a continuously Fréchet differentiable (and hence continuous) mapping on X. This is equivalent to the statement that at every $x \in X$ there exists the Jacobian matrix

$$J_f(x) = \begin{pmatrix} f_{1x_1} & \cdots & f_{1x_k} \\ \vdots & \ddots & \vdots \\ f_{kx_1} & \cdots & f_{kx_k} \end{pmatrix}, \qquad x = (x_1, \ldots, x_k),$$

and depends continuously on x.

The Fréchet derivative is then given by

$$f'_x(h) = J_f(x)\, h, \qquad h \in \mathbb{R}^k,$$

and we obtain

$$||f'_x|| = ||J_f(x)|| = \sup\{||J_f(x)\, h|| \mid ||h|| = 1\}.$$

Theorem 1.5 now gives rise to the following

Corollary. *Let $x^* \in X$ be a fixed point of f. Then the following two statements are true:*

a) Let the spectral radius $\varrho(J_f(x^*))$ be smaller than 1. Then x^* is asymptotically stable with respect to f.
b) Let $J_f(x^*)$ be invertible and let all the eigenvalues of $J_f(x^*)$ be larger than 1 in absolute value. Then x^* is unstable with respect to f.

Proof. By Theorem 3 in Chapter 1 of the book "*Analysis of Numerical Methods*" by E. Isaacson and H.B. Keller (John Wiley and Sons, New York, London, Sydney 1966) there exists, for every $\delta > 0$, a vector norm in $\mathbb{R}^k$ such that

$$||J_f(x^*)|| \leq \varrho(J_f(x^*)) + \delta.$$

a) Let us choose

$$\delta = \frac{1}{2}(1 - \varrho(J_f(x^*))) \quad (> 0).$$

Then

$$||f'_{x^*}|| = ||J_f(x^*)|| \leq \frac{1}{2}(1 + \delta(J_f(x^*))) < 1$$

and the assertion follows from Theorem 1.5 a).

b) From the assumption it follows that $\varrho\left(J_f(x^*)^{-1}\right) < 1$. Again by the above quoted theorem it follows that, for a suitable matrix norm,

$$||J_f(x^*)^{-1}|| < 1 \quad \Longrightarrow \quad 1 < ||J_f(x^*)^{-1}||^{-1} = ||{f'_{x^*}}^{-1}||^{-1}$$

so that the assertion follows from Theorem 1.5 b). □

Assumption a) in the Corollary is not necessary for the asymptotic stability of x^* with respect to f as can be seen by the Example (1.45).

Here we have

$$f_{1x}(x,y) = -\frac{2xy}{(1+x^2)^2}, \qquad f_{1y}(x,y) = \frac{1}{1+x^2},$$
$$f_{2x}(x,y) = \frac{1}{1+y^2}, \qquad f_{2y}(x,y) = -\frac{2xy}{(1+y^2)^2},$$

hence

$$J_f(0,0) = \begin{pmatrix} 0 & 1 \\ 1 & 0 \end{pmatrix} \quad \Longrightarrow \quad \varrho\left(J_f(0,0)\right) = 1.$$

However, we have seen in Section 1.3.3 that $(0,0)$ is asymptotically stable with respect to f.

1.3.5 Linear Systems

As at the beginning of Section 1.3.4 we consider a normed linear space $(E, ||\cdot||)$ and a mapping $f : E \to E$ which is given by

$$f(x) = A(x) + b, \qquad x \in E,$$

where $A : E \to E$ is a continuous linear mapping and $b \in E$ a fixed element. Then f is Fréchet differentiable at every $x \in E$ and its Fréchet derivative is given by

$$f'_x = A \qquad \text{for all} \quad x \in E.$$

Theorem 1.5 leads immediately to

Theorem 1.6. *Let $x^* \in E$ be a fixed point of f, i.e.*

$$x^* = A\left(x^*\right) + b. \tag{1.49}$$

Then the following two statements are true:

a) If $||A|| < 1$, then x^* is asymptotically stable with respect to f.
b) Let A be continuously invertible and $||A^{-1}||^{-1} > 1$. Then x^* is unstable with respect to f.

Exercise 1.5. Show that under the assumption $||A|| < 1$ there is at most one $x^* \in E$ with (1.49).

Further show that, if $||A|| < 1$ and there exists $x^* \in E$ with (1.49) (which is then unique), it follows that, for every $x_0 \in E$, the sequence $(x_n)_{n\in\mathbb{N}_0}$ given by $x_{n+1} = A\left(x_n\right) + b$, $n \in \mathbb{N}_0$, converges to x^*.

The Corollary of Theorem 1.5 leads to the following

Corollary. *Let $E = \mathbb{R}^k$ equipped with any norm and let*

$$f(x) = Ax + b, \qquad x \in \mathbb{R}^k,$$

where A is a real $k \times k-$matrix and $b \in \mathbb{R}^k$ a fixed element.

Let $x^* \in \mathbb{R}^k$ be a fixed point of f, i.e.

$$x^* = Ax^* + b.$$

Then the following two statements are true:

a) If $\varrho(A) < 1$, then x^* is asymptotically stable with respect to f.
b) If A is invertible and all eigenvalues of A are larger than 1 in absolute value, then x^* is unstable with respect to f.

In the following we introduce a concept of stability and asymptotic stability for linear systems that differ from the one given in Section 1.3.3 and that we adopt from [23]. For that purpose we consider the sequences $(x_n)_{n\in\mathbb{N}_0}$ in E with

$$x_{n+1} = Ax_n + b, \qquad n \in \mathbb{N}_0, \; x_0 \in E. \tag{1.50}$$

Definition. *A sequence $(x_n)_{n\in\mathbb{N}_0}$ in E with (1.50) is called*

1. *stable, if for every $\varepsilon > 0$ and every $N \in \mathbb{N}$, there exists some $\delta = \delta(\varepsilon, N)$ such that for every sequence $(\overline{x}_n)_{n\in\mathbb{N}_0}$ with (1.50) for $\overline{x}_0 \in E$ and $||\overline{x}_N - x_N|| < \delta$ it follows that*

$$||\overline{x}_n - x_n|| < \varepsilon \qquad \text{for all} \quad n \geq N,$$

2. *attractive, if for every $N \in \mathbb{N}$, there exists some $\delta = \delta(N)$ such that for every sequence $(\overline{x}_n)_{n\in\mathbb{N}_0}$ with (1.50) for $\overline{x}_0 \in E$ and $||\overline{x}_N - x_N|| < \delta$ it follows that*

$$\lim_{n\to\infty} ||\overline{x}_n - x_n|| = 0,$$

3. *asymptotically stable, if $(x_n)_{n\in\mathbb{N}_0}$ is stable and attractive.*

In order to guarantee the stability of all sequences $(x_n)_{n\in\mathbb{N}_0}$ with (1.50) it suffices to guarantee the stability of the zero sequence $(x_n = \Theta_E)_{n\in\mathbb{N}_0}$ which satisfies (1.50) for $b = \Theta_E$ =zero element of E. This is a consequence of

Lemma 1.3. *The following statements are equivalent:*

(1) All sequences $(x_n)_{n\in\mathbb{N}_0}$ with (1.50) are stable.
(2) One sequence $(x_n)_{n\in\mathbb{N}_0}$ with (1.50) is stable.
(3) The zero sequence $(x_n = \Theta_E)_{n\in\mathbb{N}_0}$ which satisfies (1.50) for $b = \Theta_E$ is stable.

Proof. (1) $\Longrightarrow$ (2) is trivially true. Now let (2) be true. Then there is a sequence $(y_n)_{n\in\mathbb{N}_0}$ with (1.50) which is stable. Let $(\overline{x}_n)_{n\in\mathbb{N}_0}$ be any sequence with (1.50) for $b = \Theta_E$. Then $(\overline{x}_n + y_n)_{n\in\mathbb{N}_0}$ satisfies (1.50) and for every $\varepsilon > 0$ there exists some $\delta = \delta(\varepsilon, N)$ such that

$$||\overline{x}_N - \Theta_E|| = ||\overline{x}_N + y_N - y_N|| < \delta \qquad \text{implies}$$

$$||\overline{x}_n - \Theta_E|| = ||\overline{x}_n + y_n - y_n|| < \varepsilon \qquad \text{for all} \quad n \geq N,$$

since $(y_n)_{n\in\mathbb{N}_0}$ is stable. From this it follows that $(x_n = \Theta_E)_{n\in\mathbb{N}_0}$ is stable which shows (2) $\Longrightarrow$ (3).

Now let $(y_n)_{n\in\mathbb{N}_0}$ and $(\overline{y}_n)_{n\in\mathbb{N}_0}$ be arbitrary sequences with (1.50). Then we choose any sequence $(z_n)_{n\in\mathbb{N}_0}$ in E with (1.50) and define

$$x_n = y_n - z_n \quad \text{and} \quad \overline{x}_n = \overline{y}_n - z \qquad \text{for} \quad n \in \mathbb{N}_0.$$

Since according to (3) the zero sequence $(\widetilde{x}_n = \Theta_E)_{n\in\mathbb{N}_0}$ (which satisfies (1.50) for $b = \Theta_E$) is stable, it follows that for every $\varepsilon > 0$ there exists some $\delta > \delta(\varepsilon, N)$ such that

$$||y_N - \overline{y}_N|| = ||x_N - \overline{x}_N - \Theta_E|| < \delta \qquad \text{implies that}$$

$$||y_n - \overline{y}_n|| = ||x_n - \overline{x}_n - \Theta_E|| < \varepsilon \qquad \text{for all} \quad n \geq N$$

which shows that (1) is true.

Hence (1) $\Longrightarrow$ (2) $\Longrightarrow$ (3) $\Longrightarrow$ (1) which completes the proof. □

Remark. *Lemma 1.3 also holds true, if we replace stable by attractive. Hence it is also true with asymptotically stable instead of stable.*

Now let us again consider the special case $E = \mathbb{R}^k$ equipped with any norm $||\cdot||$. According to Lemma 1.3 we consider sequences $(x_n)_{n\in\mathbb{N}_0}$ in $\mathbb{R}^k$ with

$$x_{n+1} = A x_n, \qquad n \in \mathbb{N}_0,\ x_0 \in \mathbb{R}^k. \tag{1.51}$$

In order to show stability of the zero sequence $(x_n = \Theta_k)_{n\in\mathbb{N}_0}$ we assume that A has eigenvalues $\lambda_1, \ldots, \lambda_k \in \mathbb{C}$ such that there exist eigenvectors $v_1, \ldots, v_k \in \mathbb{C}^k$ which are linearly independent. Therefore every $x_0 \in \mathbb{R}^k$ has a unique representation

$$x_0 = \sum_{i=1}^{k} c_i\, v_i, \qquad c_1, \ldots, c_k \in \mathbb{C}.$$

This implies that every sequence $(x_n)_{n\in\mathbb{N}_0}$ can be represented in the form

$$x_n = \sum_{i=1}^{k} c_i \lambda_i^n\, v_i, \qquad n \in \mathbb{N}_0.$$

Now let us assume that

$$|\lambda_i| \leq 1 \qquad \text{for} \quad i = 1, \ldots, k. \tag{1.52}$$

If we define, for every

$$z = \sum_{i=1}^{k} c_i(z)\, v_i \in \mathbb{C}^k,$$

a norm by

$$||z|| = \sum_{i=1}^{k} |c_i(z)|,$$

it follows that

$$\begin{aligned} ||x_n|| &= \sum_{i=1}^{k} |\lambda_i|^n \, |c_i\,(x_0)\,| \\ &\leq \sum_{i=1}^{k} |c_i\,(x_0)\,| = ||x_0||. \end{aligned}$$

This leads to

Theorem 1.7. *Let the eigenvalues $\lambda_1, \ldots, \lambda_k \in \mathbb{C}$ of A satisfy (1.52) and be such that the corresponding eigenvectors are linearly independent. Then the zero sequence $(x_n = \Theta_k)_{n \in \mathbb{N}_0}$ which satisfies (1.51) is stable and hence every sequence $(x_n)_{n \in \mathbb{N}_0}$ that satisfies (1.51) with $E = \mathbb{R}^k$ is stable.*

Proof. Let $\varepsilon > 0$ be chosen. Then we put $\delta = \varepsilon$. Now let $(\overline{x}_n)_{n \in \mathbb{N}_0}$ be any sequence with (1.51) for $x_n = \overline{x}_n$ and $||\overline{x}_0|| < \delta$. Then it follows that $||\overline{x}_n - \Theta_k|| < \varepsilon$ for all $n \in \mathbb{N}_0$ which completes the proof. □

Exercise 1.6. Prove that the zero sequence $(x_n = \Theta_k)_{n \in \mathbb{N}_0}$ is asymptotically stable, if all eigenvalues of A are less than 1 in absolute value.

Remark. *Under the assumption of Theorem 1.7 it is also true that the mapping $f : \mathbb{R}^k \to \mathbb{R}^k$ defined by*

$$f(x) = A\,x, \qquad x \in \mathbb{R}^k,$$

has $\{\Theta_k\}$ as a stable fixed point in the sense of the definition at the beginning of Section 1.3.3.

Proof. Let $G = \mathbb{R}^k$ and define $V : \mathbb{R}^k \to \mathbb{R}$ by

$$V(x) = ||x||, \qquad x \in \mathbb{R}^k,$$

where

$$||x|| = \sum_{i=1}^{k} |c_i(x)|, \qquad x \in \mathbb{R}^k.$$

Then it follows that

$$V(Ax) = ||Ax|| = \sum_{i=1}^{k} |\lambda_i|\, |c_i(x)| \le \sum_{i=1}^{k} |c_i(x)| = ||x||, \qquad x \in \mathbb{R}^k.$$

Hence $V(Ax) - V(x) \le 0$ for all $x \in G$ and V is a Lyapunov function on $G = \mathbb{R}^k$.

Further we have

$$V(x) \ge 0 \quad \text{for all} \quad x \in G \qquad \text{and} \qquad (V(x) = 0 \iff x = \Theta_k).$$

Theorem 1.3 therefore implies that $\{\Theta_k\}$ is a stable fixed point of f. □

We can also use another Lyapunov function in order to show that $\widehat{x} = \Theta_k$ is a stable fixed point of the system (1.51). For that purpose we choose a symmetric and positive definite real $k \times k-$matrix B and define a function

$$V(x) = x^T B\, x, \qquad x \in \mathbb{R}^k.$$

If

$$x^T \left(A^T B\, A - B\right) x \le 0 \qquad \text{for all} \quad x \in \mathbb{R}^k,$$

then V is a Lyapunov function with respect to

$$f(x) = A\,x, \qquad x \in \mathbb{R}^k,$$

on $G = \mathbb{R}^k$ which is positive definite with respect to $\{\Theta_k\}$.

Exercise 1.7.

a) Show with the aid of Theorem 1.3 that $\{\Theta_k\}$ is stable with respect to f.
b) Show with the aid of Theorem 1.3 that $\{\Theta_k\}$ is asymptotically stable with respect to f, if

$$x^T \left(A^T B\, A - B\right) x < 0 \qquad \text{for all} \quad x \in \mathbb{R}^k,\ x \ne \Theta_k.$$

One can even show that $\{\Theta_k\}$ is globally asymptotically stable with respect to f, i.e., $\{\Theta_k\}$ is stable with respect to f and

$$A^n x_0 \to \Theta_k \quad \text{as} \quad n \to \infty \qquad \text{for all} \quad x_0 \in \mathbb{R}^k$$

which is equivalent to

$$A^n \to 0 = k \times k - \text{zero matrix}.$$

Conversely, let $\{\Theta_k\}$ be globally asymptotically stable with respect to f. Further let C be a symmetric and positive definite real $k \times k$–matrix such that the series $\sum_{k=0}^{\infty} \left(A^T\right)^k C\, A^k$ converges. If we define

$$B = \sum_{k=0}^{\infty} \left(A^T\right)^k C\, A^k,$$

then B is a symmetric and positive definite real $k \times k$–matrix and it follows that

$$A^T B\, A - B = -C$$

which implies

$$x^T \left(A^T B\, A - B\right) x < 0 \qquad \text{for all} \quad x \in \mathbb{R}^k, \; x \neq \Theta_k.$$

1.3.6 Discretization of Time-Continuous Dynamical Systems

As in Section 1.1 we consider an autonomous system of the form

$$\dot{x} = f(x) \tag{1.17}$$

where $f \in C^1(\mathbb{R}^n, \mathbb{R}^n)$. Then for every $x \in \mathbb{R}^n$ there is exactly one solution $\varphi = \varphi(x,t) \in (-\alpha, \alpha)$ for some $\alpha > 0$ with $\varphi \in C^1((-\alpha,\alpha), \mathbb{R}^n)$ and

$$\varphi(x,0) = x.$$

If we assume that $\alpha = \infty$, then by the definition

$$\pi(x,t) = \varphi(x,t) \qquad \text{for all} \quad x \in \mathbb{R}^n \quad \text{and} \quad t \in \mathbb{R}$$

a time-continuous dynamical system $\pi : \mathbb{R}^n \times \mathbb{R} \to \mathbb{R}^n$ is defined as being shown in Section 1.1. For a given $h > 0$ we now replace the derivatives $\dot{x}_i = \dot{x}_i(t)$ on the left-hand side of (1.17) by difference quotients

$$\frac{x_i(t+h) - x_i(t)}{h}, \qquad i = 1, \ldots, n,$$

and obtain a system of difference equations of the form

$$x_i(t+h) = x_i(t) + h\, f_i(x(t)), \qquad t \in \mathbb{R}, \; i = 1, \ldots, n.$$

If one defines a vector function $g^h : \mathbb{R}^n \to \mathbb{R}^n$ by

$$g_i^h(x) = x_i + h\, f_i(x(t)), \qquad x \in \mathbb{R}^n,\ i = 1, \dots, n, \tag{1.53}$$

then g^h is continuous and with the definition

$$\begin{aligned} \pi_h(x,k) &= \left(g^h\right)^k(x) = \underbrace{g^h \circ g^h \circ \dots \circ g^h}_{k-\text{times}}(x) \qquad \text{for} \quad x \in \mathbb{R}^n,\ k \in \mathbb{N}, \\ \pi_h(x,0) &= x, \qquad x \in \mathbb{R}^n, \end{aligned} \tag{1.54}$$

we obtain a time-discrete dynamical system (see Section 1.3.1) which is called a discretization of (1.17) (with stepsize h).

A point $\widehat{x} \in \mathbb{R}^n$ is a rest point of the system (1.17), i.e., a solution of the equation $f(\widehat{x}) = \Theta_n$, if and only if $\widehat{x}$ is a fixed point of g^h, i.e., a solution of the equation $g^h(\widehat{x}) = \widehat{x}$.

In Section 1.2.3 we have shown (see Proposition 1.13) that a rest point $x \in \mathbb{R}^n$ of the system (1.17) is asymptotically stable, if the Jacobi matrix

$$J_f(\widehat{x}) = \left(f_{i x_j}(\widehat{x})\right)_{i,j=1,\dots,n}$$

of f at $\widehat{x}$ has only eigenvalues with negative real parts.

The Jacobi matrix of g^h in $\widehat{x} \in \mathbb{R}^n$ reads

$$J_{g^h}(\widehat{x}) = \left(\delta_{ij} + h\, f_{i x_j}(\widehat{x})\right)_{i,j=1,\dots,n}$$

where

$$\delta_{ij} = \begin{cases} 0 & \text{for} \quad i \neq j, \\ 1 & \text{for} \quad i = j. \end{cases}$$

Obviously $\lambda \in \mathbb{C}$ is an eigenvalue of $J_f(\widehat{x})$, if and only if $1+\lambda\, h$ is an eigenvalue of $J_{g^h}(\widehat{x})$. Further we have

$$\begin{aligned} |1 + \lambda\, h|^2 &= (1 + \mathrm{Re}(\lambda)\, h)^2 + (\mathrm{Im}(\lambda)\, h)^2 \\ &= 1 + 2\mathrm{Re}(\lambda)\, h + h^2\, |\lambda|^2 \\ &= 1 + h\left(2\mathrm{Re}(\lambda) + h\, |\lambda|^2\right). \end{aligned}$$

This equation implies that $\mathrm{Re}(\lambda) < 0$, if and only if $|1+\lambda\, h| < 1$ for sufficiently small $h > 0$. We can even say that

$$|1 + \lambda\, h|^2 < 1 \qquad \Longleftrightarrow \qquad \mathrm{Re}(\lambda) < 0 \quad \text{and} \quad h < \frac{2\,\mathrm{Re}(\lambda)}{|\lambda|^2}. \tag{1.55}$$

This result and the Corollary of Theorem 1.5 then lead to the following

Theorem 1.8. *Let $\widehat{x} \in \mathbb{R}^n$ be a rest point of the system (1.17). Then the following two statements are true:*

a) If the Jacobi matrix $J_f(\widehat{x})$ of f in $\widehat{x}$ has only eigenvalues with negative real parts, then, for sufficiently small $h > 0$, $\widehat{x}$ is an asymptotically stable fixed point of g^h.

b) Let $J_{g^h}(\widehat{x})$ be invertible and let all the eigenvalues of $J_{g^h}(\widehat{x})$ be larger than 1 in absolute value. Then $\widehat{x}$ is an unstable fixed point of g^h.

1.3.7 Applications

a) Discretization of a Predator-Prey Model

We consider the following predator-prey model

$$\begin{aligned} \dot{x}_1(t) &= (c_1 + c_{12}\, x_2(t))\ x_1(t), \\ \dot{x}_2(t) &= (c_2 + c_{21}\, x_1(t) + c_{22}\, x_2(t))\ x_2(t), \qquad t \in \mathbb{R}_+, \end{aligned} \tag{1.56}$$

where $c_1 < 0,\ c_2 > 0,\ c_{12} > 0,\ c_{21} < 0,\ c_{22} \le 0$ (see (1.23)).

We assume that

$$c_2 c_{12} - c_1 c_{22} > 0.$$

Then this system has exactly one rest point $\widehat{x} = (\widehat{x}_1, \widehat{x}_2)$, $\widehat{x}_1 > 0,\ \widehat{x}_2 > 0$ which is given by

$$\widehat{x}_1 = \frac{c_2 c_{12} - c_1 c_{22}}{-c_{12} c_{21}}, \qquad \widehat{x}_2 = -\frac{c_1}{c_{12}}. \tag{1.57}$$

Let

$$X = \left\{ x \in \mathbb{R}^2 \mid x_1 > 0 \text{ and } x_2 > 0 \right\}.$$

Then it follows (see Section 1.2.4) that $x \in \mathcal{C}(\mathbb{R}_+, X)$ is a solution of the system (1.56), if and only if $u \in \mathcal{C}^1\left(\mathbb{R}_+, \mathbb{R}^2\right)$ with

$$u_1(t) = \ln\left(\frac{x_1(t)}{\widehat{x}_1}\right) \quad \text{and} \quad u_2(t) = \ln\left(\frac{x_2(t)}{\widehat{x}_2}\right), \qquad t \in \mathbb{R}_+,$$

is a solution of the system

$$\begin{aligned} \dot{u}_1(t) &= c_{12}\, \widehat{x}_2 \left(\mathrm{e}^{u_2(t)} - 1\right), \\ \dot{u}_2(t) &= c_{21}\, \widehat{x}_1 \left(\mathrm{e}^{u_1(t)} - 1\right) + c_{22}\, \widehat{x}_2 \left(\mathrm{e}^{u_2(t)} - 1\right), \qquad t \in \mathbb{R}_+. \end{aligned} \tag{1.58}$$

The only rest point of this system is the point $(0,0)$. Further it follows that $(0,0)$ is asymptotically stable, if and only if $\widehat{x} = (\widehat{x}_1, \widehat{x}_2)$ given by (1.57) is asymptotically stable.

If we define a vector function $f : \mathbb{R}^2 \to \mathbb{R}^2$ by

$$f(u_1, u_2) = \begin{pmatrix} c_{12}\,\widehat{x}_2\,\left(e^{u_2(t)} - 1\right) \\ c_{21}\,\widehat{x}_1\,\left(e^{u_1(t)} - 1\right) + c_{22}\,\widehat{x}_2\,\left(e^{u_2(t)} - 1\right) \end{pmatrix}, \qquad (u_1, u_2) \in \mathbb{R}^2, \tag{1.59}$$

then the system (1.58) can be written in the form

$$\dot{u}(t) = f(u(t)), \qquad t \in \mathbb{R}_+,$$

and the Jacobi matrix of f at $(0,0)$ is given by

$$J_f(0,0) = \begin{pmatrix} 0 & c_{12}\,\widehat{x}_2 \\ c_{21}\,\widehat{x}_1 & c_{22}\,\widehat{x}_2 \end{pmatrix}.$$

The eigenvalues of $J_f(0,0)$ are given by

$$\lambda_{1,2} = \frac{1}{2}\,c_{22}\,\widehat{x}_2 \pm \sqrt{\frac{1}{4}\,c_{22}^2\,\widehat{x}_2^2 + c_{12}c_{21}\,\widehat{x}_1\widehat{x}_2}.$$

We now distinguish two cases:

a) Let $c_{22} < 0$. Then it follows that

$$\mathrm{Re}\,(\lambda_1) < 0 \qquad \text{and} \qquad \mathrm{Re}\,(\lambda_2) < 0.$$

Theorem 1.8 then implies that $(0,0)$ is an asymptotically stable fixed point of the mapping $g^h : \mathbb{R}^2 \to \mathbb{R}^2$ defined by (1.53) with f given by (1.59), if $h > 0$ is sufficiently small. From the equivalence (1.55) we deduce that this is the case, if

$$h < \min\left(\frac{2\,\mathrm{Re}\,(\lambda_1)}{|\lambda_1|^2}, \frac{2\,\mathrm{Re}\,(\lambda_2)}{|\lambda_2|^2}\right).$$

b) Let $c_{22} = 0$. Then it follows that

$$\lambda_{1,2} = \pm i\sqrt{|c_{12}c_{21}|\,\widehat{x}_1\widehat{x}_2}, \qquad i = \sqrt{-1}.$$

Hence

$$\mathrm{Re}\,(\lambda_1) = \mathrm{Re}\,(\lambda_2) = 0$$

and the absolute values of the eigenvalues of the Jacobi matrix $J_{g^h}(0,0)$ of g^h at $(0,0)$ are given by

$$|1 + \lambda_1\,h| = \sqrt{1 + h^2\,|\lambda_1|^2} \qquad \text{and} \qquad |1 + \lambda_2\,h| = \sqrt{1 + h^2\,|\lambda_2|^2}.$$

This implies that they are larger than 1 for every choice of $h > 0$.

Theorem 1.8 then implies that $(0,0)$ is an unstable fixed point of g^h for every choice of $h > 0$.

b) Discretization of a Model of Interacting Logistic Growth of Two Populations

The model is described by the system

$$\begin{aligned} \dot{x}_1(t) &= (c_1 + c_{11}\,x_1(t) + c_{12}\,x_2(t))\;x_1(t), \\ \dot{x}_2(t) &= (c_2 + c_{21}\,x_1(t) + c_{22}\,x_2(t))\;x_2(t), \qquad t \in \mathbb{R}_+, \end{aligned} \tag{1.60}$$

where $c_1 > 0,\; c_2 > 0,\; c_{11} < 0,\; c_{12} < 0,\; c_{21} < 0,\; c_{22} < 0.$

If we define

$$f(x) = \begin{pmatrix} (c_1 + c_{11}\,x_1 + c_{12}\,x_2)\;x_1 \\ (c_2 + c_{21}\,x_1 + c_{22}\,x_2)\;x_2 \end{pmatrix}, \qquad x = (x_1, x_2)\,, \tag{1.61}$$

then the system (1.60) can be written in the form

$$\dot{x}(t) = f(x(t)), \qquad t \in \mathbb{R}_+.$$

A point $\widehat{x} \in \mathbb{R}^2$ with $\widehat{x}_1 > 0$ and $\widehat{x}_2 > 0$ is a rest point of the system (1.60), i.e., a solution of the equation $f\left(\widehat{x}\right) = \Theta_2$, if and only if

$$\begin{aligned} c_{11}\,\widehat{x}_1 + c_{12}\,\widehat{x}_2 &= -c_1, \\ c_{21}\,\widehat{x}_1 + c_{22}\,\widehat{x}_2 &= -c_2. \end{aligned} \tag{1.62}$$

Let us assume that

$$c_{11}c_{22} - c_{12}c_{21} > 0. \tag{1.63}$$

Then the equation (1.62) has exactly one solution $\widehat{x} = (\widehat{x}_1, \widehat{x}_2)$ which is given by

$$\widehat{x}_1 = \frac{-c_1 c_{22} + c_2 c_{12}}{c_{11}c_{22} - c_{12}c_{21}}, \qquad x_2 = \frac{c_1 c_{21} - c_2 c_{11}}{c_{11}c_{22} - c_{12}c_{21}}.$$

Further it follows that $\widehat{x}_1 > 0$ and $\widehat{x}_2 > 0$, if and only if

$$-c_1 c_{22} + c_2 c_{12} > 0 \qquad \text{and} \qquad c_1 c_{21} - c_2 c_{11} > 0. \tag{1.64}$$

The Jacobi matrix of f at $\widehat{x}$ is given by

$$J_f\left(\widehat{x}\right) = \begin{pmatrix} c_{11}\,\widehat{x}_1 & c_{12}\,\widehat{x}_1 \\ c_{21}\,\widehat{x}_2 & c_{22}\,\widehat{x}_2 \end{pmatrix}$$

and has the eigenvalues

$$\lambda_{1,2} = \frac{1}{2}\left(c_{11}\,\widehat{x}_1 + c_{22}\,\widehat{x}_2\right) \pm \sqrt{\frac{1}{4}\left(c_{11}\,\widehat{x}_1 + c_{22}\,\widehat{x}_2\right) + \left(c_{12}c_{21} - c_{11}c_{22}\right)\,\widehat{x}_1\widehat{x}_2}$$

which have negative real parts.

Theorem 1.8 then implies that $\widehat{x}$ is an asymptotically stable fixed point of the mapping $g^h : \mathbb{R}^2 \to \mathbb{R}^2$ defined by (1.53) with f given by (1.61), if $h > 0$ is sufficiently small.

c) An Emission Reduction Model

In [32] a mathematical model for the reduction of carbon dioxide emission is investigated in form of a time-discrete dynamical system which as uncontrolled system is given by the following system of difference equations

$$\begin{aligned} E_i(t+1) &= E_i(t) + \sum_{j=1}^{r} em_{ij}\, M_j(t), \\ M_i(t+1) &= M_i(t) - \lambda_i\, M_i(t)\,(M_i^* - M_i(t))\, E_i(t) \\ &\text{for} \quad i = 1, \ldots, r \quad \text{and} \quad t \in \mathbb{N}_0 \end{aligned} \tag{1.65}$$

where $E_i(t)$ denotes the amount of emission reduction and $M_i(t)$ the financial means spent by the $i-$th actor at the time t, $\lambda_i > 0$ is a growth parameter and $M_i^* > 0$ an upper bound for $M_i(t)$ for $i = 1, \ldots, r$ and $t \in \mathbb{N}_0$. For $t = 0$ we assume the system to be in the state $E_{0i}, M_{0i},\ i = 1, \ldots, r$ which leads to the initial conditions

$$E_i(0) = E_{0i} \quad \text{and} \quad M_i(0) = M_{0i} \qquad \text{for} \quad i = 1, \ldots, r.$$

If we define $x = \left(E^T, M^T\right)^T,\ E, M \in \mathbb{R}^r$, and functions $f_i : \mathbb{R}^n \to \mathbb{R}^n,\ i = 1, \ldots, n = 2r$ by

$$\begin{aligned} f_i(x) &= E_i(t) + \sum_{j=1}^{r} em_{ij}\, M_j, & i &= 1, \ldots, r, \\ f_i(x) &= M_i - \lambda_i\, M_i\,(M_i^* - M_i)\, E_i, & i &= r+1, \ldots, n, \end{aligned} \tag{1.66}$$

then we can write (1.65) in the form

$$x(t+1) = f(x(t)), \qquad t \in \mathbb{N}_0, \tag{1.67}$$

where $f(x) = (f_1(x), \ldots, f_n(x))^T$.

For every $\widehat{x} = \left(\widehat{E}^T, \Theta_r^T\right)^T,\ E \in \mathbb{R}^r$, we have

$$\widehat{x} = f(\widehat{x}).$$

Let $\widehat{x}$ be any such fixed point of f.

Then we replace the system (1.67) by the linear system

$$x(t+1) = J_f(\widehat{x})\, x(t), \qquad t \in \mathbb{N}_0, \tag{1.68}$$

where the Jacobi matrix $J_f(\widehat{x})$ is given by

$$J_f(\widehat{x}) = \begin{pmatrix} I_r & C \\ O_r & D \end{pmatrix}$$

where I_r and O_r is the $r \times r$–unit and –zero matrix, respectively, and

$$C = \begin{pmatrix} em_{11} & \cdots & em_{1r} \\ \vdots & \ddots & \vdots \\ em_{r1} & \cdots & em_{rr} \end{pmatrix}$$

and

$$D = \begin{pmatrix} d_{11} & & 0 \\ & \ddots & \\ 0 & & d_{rr} \end{pmatrix}$$

with

$$d_{ii} = 1 - \lambda_i \, M_i^* \widehat{E}_i, \qquad i = 1, \ldots, r.$$

Let us assume that

$$d_{ii} \neq 0, \; d_{ii} \neq 1 \quad \text{and} \quad |d_{ii}| \leq 1 \qquad \text{for all} \quad i = 1, \ldots, r.$$

Then the matrix $J_f\left(\widehat{x}\right)$ is non-singular. Its eigenvalues are given by

$$\mu_i = 1 \quad \text{and} \quad \mu_{i+r} = d_{ii} \qquad \text{for} \quad i = 1, \ldots, r,$$

hence

$$|\mu_i| \leq 1 \qquad \text{for} \quad i = 1, \ldots, 2r$$

and the corresponding eigenvectors

$$\begin{pmatrix} e_i \\ \Theta_r \end{pmatrix} \qquad \text{for} \quad i = 1, \ldots, r$$

and

$$\begin{pmatrix} -C e_j / \left(1 - d_{jj}\right) \\ e_j \end{pmatrix} \qquad \text{for} \quad j = 1, \ldots, r$$

($e_j = j -$ th unit vector) are linearly independent. Theorem 1.7 therefore implies that the zero sequence $(x(t) = \Theta_n)_{t \in \mathbb{N}_0}$ which satisfies (1.68) is stable.

1.3.8 The Non-Autonomous Case: Definitions and Elementary Properties

Let X be a metric space with metric $d : X \times X \to \mathbb{R}_+$ and let $(f_n)_{n\in\mathbb{N}}$ be a sequence of continuous mappings $f_n : X \to X$, $n \in \mathbb{N}$. Then the pair $\left(X, (f_n)_{n\in\mathbb{N}}\right)$ is called a non-autonomous time-discrete dynamical system. The dynamics in this system is defined by the sequence $F = (F_n)_{n\in\mathbb{N}_0}$ of mappings $F_n : X \to X$ given by

$$F_n(x) = f_n \circ f_{n-1} \circ \ldots \circ f_1(x) \qquad \text{for all} \quad x \in X \quad \text{and} \quad n \in \mathbb{N}$$

and

$$F_0(x) = x \qquad \text{for all} \quad x \in X.$$

If $f_n = f$ for all $n \in \mathbb{N}$ we obtain an autonomous time-discrete dynamical system (x, f) as being defined in Section 1.3.1.

For every $x \in X$ we define an orbit starting with x by

$$\gamma_F(x) = \bigcup_{n\in\mathbb{N}_0} \{F_n(x)\} \tag{1.69}$$

and

$$L_F(x) = \bigcap_{n\in\mathbb{N}_0} \overline{\bigcup_{m\geq n} \{F_m(x)\}} \tag{1.70}$$

where $\overline{A}$ denotes the closure of $A \subseteq X$.

Then we can prove

Proposition 1.19. *For every $x \in X$ the limit set $L_F(x)$ consists of all accumulation points of the sequence $(F_n(x))_{n\in\mathbb{N}_0}$.*

The proof is the same as that of Proposition 1.16 and is left as an exercise.

Now let X be compact. Then, for every $x \in X$, the sequence $(F_n(x))_{n\in\mathbb{N}_0}$ has at least one accumulation point which implies, by Proposition 1.19, that $L_F(x)$ is non-empty. By definition $L_F(x)$ is a closed subset of X and hence also compact for every $x \in X$.

Further one can show that $L_F(x)$ is the smallest closed subset $S \subseteq X$ with

$$\lim_{n\to\infty} \varrho(F_n(x), S) = 0 \qquad \text{with} \quad \varrho(y, S) = \min\{d(y, z) \mid z \in S\}. \tag{1.71}$$

The proof is left as an exercise (see Proposition 1.18). In addition we can prove the following

Proposition 1.20. *Let X be compact and let $(f_n)_{n\in\mathbb{N}}$ be uniformly convergent to some mapping $f_0 : X \to X$. Then it follows that*

$$f_0\left(L_F(x)\right) = L_F(x) \qquad \textit{for all} \quad x \in X.$$

Proof.
1) At first we show that $f_0(L_F(x)) \subseteq L_F(x),\ x \in X$.

Choose $x \in X$ and $y \in L_F(x)$ arbitrarily. Since f_0 is continuous, for every $\varepsilon > 0$, there exists some $\delta = \delta(\varepsilon, y) > 0$ such that

$$f_0(\widetilde{x}) \in U_\varepsilon(f_0(y)) \qquad \text{for all} \quad \widetilde{x} \in U_\delta(y).$$

Uniform convergence of $(f_n)_{n\in\mathbb{N}}$ to f_0 implies the existence of some $n(\varepsilon) \in \mathbb{N}$ such that

$$f_n(\widetilde{x}) \in U_\varepsilon(f_0(\widetilde{x})) \subseteq U_{2\varepsilon}(f_0(y)) \qquad \text{for all} \quad \widetilde{x} \in U_\delta(y) \quad \text{and all} \quad n \geq n(\varepsilon).$$

By Proposition 1.19 there is a subsequence $(F_{n_i}(x))_{i\in\mathbb{N}_0}$ of $(F_n(x))_{n\in\mathbb{N}_0}$ with

$$F_{n_i}(x) \in U_\delta(y) \qquad \text{for all} \quad i \in \mathbb{N}_0.$$

This implies

$$F_{n_i+1}(x) = f_{n_i+1} \circ F_{n_i}(x) \in U_{2\varepsilon}(f_0(y)) \qquad \text{for all} \quad n_i \geq n(\varepsilon).$$

Since $\varepsilon > 0$ is chosen arbitrarily, it follows that

$$F_{n_i+1}(x) \to f_0(y) \qquad \text{and hence, by Proposition 1.19,} \quad f_0(y) \in L_F(x).$$

2) Next we prove $L_F(x) \subseteq f_0(L_F(x)),\ x \in X$.

Choose $\widehat{x} \in X$ and $y \in L_F(\widehat{x})$ arbitrarily. Then we have to show the existence of some $x \in L_F(\widehat{x})$ such that $y = f_0(x)$.

By Proposition 1.19 there is a subsequence $(F_{n_i}(\widehat{x}))_{i\in\mathbb{N}_0}$ of $(F_n(\widehat{x}))_{n\in\mathbb{N}_0}$ with $F_{n_i}(\widehat{x}) \to y$ as $i \to \infty$. If we put

$$x_i = F_{n_i-1}(\widehat{x}) \qquad \text{for all} \quad i \in \mathbb{N}_0,$$

then it follows that $f_{n_i}(x_i) \to y$ as $i \to \infty$. We can also assume that $x_i \to x$ for some $x \in L_F(\widehat{x})$.

Then we have

$$\begin{aligned} & d(f_0(x), y) \\ & \quad \leq d(f_0(x), f_0(x_i)) + d(f_0(x_i), f_{n_i}(x_i)) + d(f_{n_i}(x_i), y) \\ & \quad \leq \underbrace{d(f_0(x), f_0(x_i))}_{\to 0} + \underbrace{\sup_{\widetilde{x}\in X} d(f_0(\widetilde{x}), f_{n_i}(\widetilde{x}))}_{\to 0} + \underbrace{d(f_{n_i}(x_i), y)}_{\to 0} \end{aligned}$$

as $i \to \infty$, hence $y = f_0(x)$. □

Next we give a more precise localization of the limit sets $L_F(x),\ x \in X$, with the aid of a Lyapunov function which is defined as follows:

Let $G \subseteq X$ be non-empty. Then we say that $V : X \to \mathbb{R}$ is a Lyapunov function with respect to $(f_n)_{n \in \mathbb{N}}$ on G, if

(1) V is continuous on X,
(2) $V(f_n(x)) - V(x) \leq 0$ for all $x \in G$ and all $n \in \mathbb{N}$ such that $f_n(x) \in G$.

For every $c \in \mathbb{R}$ we define

$$V^{-1}(c) = \{x \in X \mid V(x) = c\}.$$

Then we can prove the following

Proposition 1.21. *Let V be a Lyapunov function with respect to $(f_n)_{n \in \mathbb{N}}$ on $G \subseteq X$ where G is relatively compact. Let further $x_0 \in G$ be chosen such that $f_n(x_0) \in G$ for all $n \in \mathbb{N}$. Then there exists some $c \in \mathbb{R}$ such that*

$$L_F(x_0) \subseteq V^{-1}(c)$$

and $L_F(x_0)$ is non-empty.

Proof. Since G is relatively compact, it follows that $L_F(x_0)$ is nonempty. For every $n \in \mathbb{N}$ we put $x_n = F_n(x_0)$ which implies

$$x_n \in G \quad \text{and} \quad V(x_{n+1}) \leq V(x_n) \qquad \text{for all} \quad n \in \mathbb{N}.$$

Since $V : X \to \mathbb{R}$ is continuous, V is bounded from below on G which implies the existence of $c = \lim_{n \to \infty} V(x_n)$.

Now let $p \in L_F(x_0)$. Then, by Proposition 1.19, there is a subsequence $(x_{n_i})_{i \in \mathbb{N}_0}$ of $(x_n)_{n \in \mathbb{N}_0}$ with $x_{n_i} \to p$ as $i \to \infty$.

This implies $V(p) = \lim_{i \to \infty} V(x_{n_i}) = c$, hence $p \in V^{-1}(c)$. □

1.3.9 Stability Based on Lyapunov's Method

Definition. *A relatively compact set $H \subseteq X$ is called stable with respect to $(f_n)_{n \in \mathbb{N}}$, if for every relatively compact open set $U \subseteq X$ with $U \supseteq \overline{H} =$ closure of H there exists an open set $W \subseteq X$ with $\overline{H} \subseteq W \subseteq U$ such that*

$$F_n(W) \subseteq U \qquad \textit{for all} \quad n \in \mathbb{N}_0$$

where

$$F_n(W) = \{F_n(x) \mid x \in W\}.$$

Theorem 1.9. *Let $H \subseteq X$ be relatively compact and such that for every relatively compact open set $U \subseteq X$ with $U \supseteq \overline{H}$ there exists an open subset B_U of U with $B_U \supseteq \overline{H}$ and*

$$f_n(B_U) \subseteq U \qquad \text{for all} \quad n \in \mathbb{N}.$$

Further let $G \subseteq X$ be an open set with $G \supseteq \overline{H}$ such that there exists a Lyapunov function V with respect to $(f_n)_{n\in N}$ on G which is positive definite with respect to $\overline{H}$, i.e.

$$V(x) \geq 0 \quad \text{for all} \quad x \in G \qquad \text{and} \qquad \left(V(x) = 0 \quad \Longleftrightarrow \quad x \in \overline{H}\,\right)$$

Then H is stable with respect to $(f_n)_{n\in N}$.

Proof. Let $U \subseteq X$ be an arbitrary relatively compact open set with $U \supseteq \overline{H}$. Then $U^* = U \cap G$ is also a relatively compact open set with $U^* \supseteq \overline{H}$ and there exists an open set $B_{U^*} \subseteq U^*$ with $B_{U^*} \supseteq \overline{H}$ and $f_n(B_{U^*}) \subseteq U^*$ for all $n \in \mathbb{N}$. Let us put

$$m = \min\left\{V(x) \mid x \in U^* \setminus B_{U^*}\right\}.$$

Since $\overline{H} \cap (U^* \setminus B_{U^*})$ is empty, it follows that $m > 0$.

If we define

$$W = \{x \in U^* \mid V(x) < m\},$$

then W is open and $\overline{H} \subseteq W \subseteq B_{U^*}$. Now let $x \in W$ be chosen arbitrarily. Then $x \in B_{U^*}$ and therefore $F_1(x) = f_1(x) \in U^*$. Further we have

$$V(F_1(x)) = V(f_1(x)) \leq V(x) < m,$$

hence $F_1(x) \in W \subseteq B_{U^*}$. This implies $F_2(x) = f_2(F_1(x)) \in U^*$ and $V(F_2(x)) \leq V(F_1(x)) < m$, hence $F_2(x) \in W$. By induction it therefore follows that

$$F_n(x) \in W \subseteq U^* \subset U \qquad \text{for all} \quad n \in \mathbb{N}_0.$$

This shows that H is stable with respect to $(f_n)_{n\in\mathbb{N}}$. □

Definition. *A set $H \subseteq X$ is called an attractor with respect to $(f_n)_{n\in\mathbb{N}}$, if there exists an open set $U \subseteq X$ with $U \supseteq \overline{H}$ such that*

$$\lim_{n\to\infty} \varrho\left(F_n(x), \overline{H}\right) = 0 \quad \left(\text{in short: } F_n(x) \to \overline{H}\,\right) \qquad \text{for all} \quad x \in U$$

where

$$\varrho\left(y, \overline{H}\right) = \inf\left\{d(y,z) \mid z \in \overline{H}\right\}.$$

If $H \subseteq X$ is stable and an attractor with respect to $(f_n)_{n\in\mathbb{N}}$, then H is called asymptotically stable with respect to $(f_n)_{n\in\mathbb{N}}$.

Theorem 1.10. *Let $H \subseteq X$ be such that there exists a relatively compact open set $U \supseteq \overline{H}$ with*

$$f_n(U) \subseteq U \qquad \text{for all} \quad n \in \mathbb{N}.$$

Further let $V : X \to \mathbb{R}$ be a Lyapunov function with respect to $(f_n)_{n\in\mathbb{N}}$ on U which is positive definite with respect to $\overline{H}$. Finally let

$$\lim_{n\to\infty} V\left(F_n(x)\right) = 0 \qquad \text{for all} \quad x \in U.$$

Then H is an attractor with respect to $(f_n)_{n\in\mathbb{N}}$.

Proof. Let $x \in U$ be chosen arbitrarily. Since the sequence $(F_n(x))_{n\in\mathbb{N}_0}$ is contained in U, for every subsequence $(F_{n_i}(x))_{i\in\mathbb{N}_0}$ of $(F_n(x))_{n\in\mathbb{N}_0}$ there exists a subsequence $\left(F_{n_{i_j}}(x)\right)_{j\in\mathbb{N}_0}$ and some $q \in \overline{U}$ with

$$\lim_{j\to\infty} F_{n_{i_j}}(x) = q.$$

This implies

$$\lim_{j\to\infty} V\left(F_{n_{i_j}}(x)\right) = V(q) = 0,$$

hence, $q \in \overline{H}$, and therefore $F_{n_{i_j}}(x) \to \overline{H}$ as $j \to \infty$. From this it follows that $F_n(x) \to \overline{H}$ which shows that H is an attractor with respect to $(f_n)_{n\in\mathbb{N}}$. □

Corollary. *Under the assumptions of Theorem 1.9 and Theorem 1.10 it follows that $H \subseteq X$ is asymptotically stable.*

Let us demonstrate Theorem 1.9 and Theorem 1.10 by an example. We choose $X = \mathbb{R}^2$ and consider a sequence $(f_n)_{n\in\mathbb{N}}$ of mappings $f_n : \mathbb{R}^2 \to \mathbb{R}^2$ given by

$$f_n(x, y) = \left(\frac{a_n y}{1 + x^2}, \frac{b_n x}{1 + y^2}\right), \qquad (x, y) \in \mathbb{R}^2, \ n \in \mathbb{N},$$

where $(a_n)_{n\in\mathbb{N}}$ and $(b_n)_{n\in\mathbb{N}}$ are sequences of real numbers with

$$a_n^2 \leq 1 \quad \text{and} \quad b_n^2 \leq 1 \qquad \text{for all} \quad n \in \mathbb{N}. \tag{1.72}$$

Put $H = \{(0, 0)\}$. If we choose

$$V(x, y) = x^2 + y^2, \qquad (x, y) \in \mathbb{R}^2,$$

then V is a Lyapunov function on $G = \mathbb{R}^2$ with respect to $(f_n)_{n\in\mathbb{N}}$ which is positive definite with respect to $\overline{H} = \{(0, 0)\}$, since

$$\begin{aligned} V\left(f_n(x, y)\right) - V(x, y) &= \left(\frac{a_n^2}{(1 + x^2)2} - 1\right) y^2 + \left(\frac{b_n^2}{(1 + y^2)2} - 1\right) x^2 \\ &\leq \left(a_n^2 - 1\right) y^2 + \left(b_n^2 - 1\right) x^2 \\ &\leq 0 \qquad \text{for all} \quad (x, y) \in \mathbb{R}^2 \quad \text{and} \quad n \in \mathbb{N}. \end{aligned}$$

Now let $U \subseteq \mathbb{R}^2$ be relatively compact and open with $(0,0) \in U$ be given. Then there exists some $r > 0$ such that

$$B_U = \{(x,y) \in \mathbb{R}^2 \mid x^2 + y^2 < r\} \subseteq U.$$

Further it follows that

$$f_{1n}(x,y)^2 + f_{2n}(x,y)^2 \leq x^2 + y^2 < r \qquad \text{for all} \quad (x,y) \in B_U,$$

hence

$$f_n(B_U) \subseteq B_U \subseteq U \qquad \text{for all} \quad n \in \mathbb{N}.$$

Therefore all the assumptions of Theorem 1.9 hold true which implies that $\{(0,0)\}$ is stable with respect to $(f_n)_{n \in \mathbb{N}}$.

Next we sharpen (1.72) to

$$a_n^2 \leq \gamma < 1 \quad \text{and} \quad b_n^2 \leq \gamma < 1 \qquad \text{for all} \quad n \in \mathbb{N}.$$

Then it follows that

$$V(f_n(x,y)) \leq \gamma \, V(x,y) \qquad \text{for all} \quad n \in \mathbb{N} \quad \text{and} \quad (x,y) \in \mathbb{R}^2$$

which implies

$$V(F_n(x,y)) \leq \gamma^n \, V(x,y) \qquad \text{for all} \quad n \in \mathbb{N} \quad \text{and} \quad (x,y) \in \mathbb{R}^2$$

and in turn

$$\lim_{n \to \infty} V(F_n(x,y)) = 0 \qquad \text{for all} \quad (x,y) \in \mathbb{R}^2.$$

By Theorem 1.10, $H = \{(0,0)\}$ is an attractor with respect to $(f_n)_{n \in \mathbb{N}}$ and by the Corollary of Theorems 1.9 and 1.10 $\{(0,0)\}$ is asymptotically stable with respect to $(f_n)_{n \in \mathbb{N}}$.

1.3.10 Linear Systems

We consider a normed linear space $(E, ||\cdot||)$ and a sequence $(f_n)_{n \in \mathbb{N}}$ of mappings $f_n : E \to E$ which are given by

$$f_n(x) = A_n(x) + b_n, \qquad x \in E, \ n \in \mathbb{N},$$

where $(A_n)_{n \in \mathbb{N}}$ is a sequence of continuous linear mappings $A_n : E \to E$ and $(b_n)_{n \in \mathbb{N}}$ is a fixed sequence in E.

Then the pair $\left(E,(f_n)_{n\in\mathbb{N}}\right)$ is a non-autonomous time-discrete dynamical system. The dynamics in this system is defined by the sequence $(F_n)_{n\in\mathbb{N}_0}$ of mappings $F_n : E \to E$ given by

$$\begin{aligned} F_n(x) &= f_n \circ f_{n-1} \circ \ldots \circ f_1(x) \\ &= A_n \circ A_{n-1} \circ \ldots \circ A_1(x) + \sum_{k=1}^{n} A_n \circ A_{n-1} \circ \ldots \circ A_{k+1}\left(b_k\right) \quad (1.73) \\ &\quad \text{for} \quad x \in E \quad \text{and} \quad n \in \mathbb{N}. \end{aligned}$$

where $A_n \circ A_{n+1} = \mathrm{id}_E$ for all $n \in \mathbb{N}$ and

$$F_0(x) = x \qquad \text{for all} \quad x \in E. \tag{1.74}$$

In general there will be no common fixed point of all f_n, i.e., no point $x^* \in E$ with $x^* = f_n(x^*)$ for all $n \in \mathbb{N}$ which then would also be a common fixed point of all F_n. Therefore fixed point stability is not a reasonable concept in this case. We replace it by another concept of stability which has been introduced in Section 1.3.5 already and whose definition will be repeated here.

Definition. *A sequence* $\left(x_n = F_n(x_0)\right)_{n\in\mathbb{N}_0}$, $x_0 \in E$, *is called*

1. *stable, if for every* $\varepsilon > 0$ *and every* $N \in \mathbb{N}_0$ *there exists some* $\delta = \delta(\varepsilon, N) > 0$ *such that for every sequence* $\left(\overline{x}_n = F_n(\overline{x}_0)\right)_{n\in\mathbb{N}_0}$, $\overline{x}_0 \in E$, *with* $||\overline{x}_N - x_N|| < \delta$ *it follows that*
$$||\overline{x}_n - x_n|| < \varepsilon \qquad \textit{for all} \quad n \geq N+1,$$
2. *attractive, if for every* $N \in \mathbb{N}_0$ *there exists some* $\delta = \delta(N) > 0$ *such that for every sequence* $\left(\overline{x}_n = F_n(\overline{x}_0)\right)_{n\in\mathbb{N}_0}$, $\overline{x}_0 \in E$, *with* $||\overline{x}_N - x_N|| < \delta$ *it follows that*
$$\lim_{n\to\infty} ||\overline{x}_n - x_n|| = 0,$$
3. *asymptotically stable, if* $\left(x_n = F_n(x_0)\right)_{n\in\mathbb{N}_0}$, $x_0 \in E$, *is stable and attractive.*

As a first consequence of this definition we have

Lemma 1.4. *The following statements are equivalent:*

(1) All sequences $\left(x_n = F_n(x_0)\right)_{n\in\mathbb{N}_0}$, $x_0 \in E$, *are stable.*
(2) One sequence $\left(x_n = F_n(x_0)\right)_{n\in\mathbb{N}_0}$, $x_0 \in E$, *is stable.*
(3) The sequence $\left(x_n = A_n \circ A_{n-1} \circ \ldots \circ A_1(\Theta_E)\right)_{n\in\mathbb{N}_0} = \left(x_n = \Theta_E\right)_{n\in\mathbb{N}_0}$ *is stable.*

The proof is the same as that of Lemma 1.3. We can also replace *stable* by *attractive* and hence by *asymptotically stable*. This Lemma leads to the following sufficient conditions for stability and asymptotic stability:

Theorem 1.11. *If*

$$||A_n|| = \sup_{||x||=1} A_n(x) \leq 1 \qquad \text{for all} \quad n \in \mathbb{N}, \tag{1.75}$$

then all sequences $(x_n = F_n(x_0))_{n\in\mathbb{N}_0}$, $x_0 \in E$, *are stable.*

If

$$\sup_{n\in\mathbb{N}} ||A_n|| < 1, \tag{1.76}$$

then all sequences $(x_n = F_n(x_0))_{n\in\mathbb{N}_0}$, $x_0 \in E$, *are asymptotically stable.*

Proof. Let us first assume that (1.75) holds true. Let $\varepsilon > 0$ be chosen arbitrarily. Then we put $\delta = \varepsilon$ and conclude that for every sequence $(x_n = F_n(x_0))_{n\in\mathbb{N}_0}$, $x_0 \in E$, with $||x_N|| < \delta$ for some $N \in \mathbb{N}_0$ it follows that

$$\begin{aligned} ||x_n - \Theta_E|| &= ||A_n \circ A_{n-1} \circ \ldots \circ A_{N+1}(x_{N+1})|| \\ &\leq ||A_n|| \cdot ||A_{n-1}|| \cdot \ldots \cdot ||A_{N+1}|| \cdot ||x_N|| \\ &< \delta = \varepsilon \qquad \text{for all} \quad n \geq N+1. \end{aligned}$$

Hence the sequence $(x_n = \Theta_E)_{n\in\mathbb{N}_0}$ is stable, and by Lemma 1.2 all sequences $(x_n = F_n(x_0))_{n\in\mathbb{N}_0}$, $x_0 \in E$ are stable.

Next we assume that (1.76) ($\Rightarrow$ (1.75)) holds true. Thus, for every $N \in \mathbb{N}_0$ there exists some $\delta = \delta(N) > 0$ such that for every sequence $(x_n = F_n(x_0))_{n\in\mathbb{N}_0}$, $x_0 \in E$, with $||x_N|| < \delta$ for some $N \in \mathbb{N}_0$ it follows that

$$\begin{aligned} \lim_{n\to\infty} ||x_n - \Theta_E|| &= \lim_{n\to\infty} ||A_n \circ A_{n-1} \circ \ldots \circ A_{N+1}(x_N)|| \\ &\leq \lim_{n\to\infty} \left(\sup_{m\in\mathbb{N}} ||A_m|| \right)^{n-N-1} ||x_N|| \\ &= 0. \end{aligned}$$

Thus, the sequence $(x_n = \Theta_E)_{n\in\mathbb{N}_0}$ is attractive, hence asymptotically stable which implies that all sequences $(x_n = F_n(x_0))_{n\in\mathbb{N}_0}$, $x_0 \in E$, are asymptotically stable.

According to the above considerations it suffices to consider homogeneous systems with

$$f_n(x) = A_n(x), \qquad x \in E, \; n \in \mathbb{N},$$

such that

$$F_n(x) = A_n \circ A_{n-1} \circ \ldots \circ A_1(x), \qquad x \in E, \; n \in \mathbb{N}.$$

Then all sequences $(x_n = F_n(x_0))_{n\in\mathbb{N}_0}$, $x_0 \in E$, are stable or asymptotically stable, if and only if the sequence $(x_n = F_n(\Theta_E))_{n\in\mathbb{N}_0}$ is stable or asymptotically stable. □

This leads to

Theorem 1.12. *Assumption: All* A_n, $n \in \mathbb{N}$, *are invertible. Assertion: The sequence* $(A_n \circ A_{n-1} \circ \ldots \circ A_1 (\Theta_E))_{n \in \mathbb{N}_0}$ *is stable, if and only if for every* $N \in \mathbb{N}_0$ *there exists a constant* $c_N > 0$ *such that*

$$||A_n \circ A_{n-1} \circ \ldots \circ A_{N+1}|| \leq c_N \qquad \text{for all} \quad n \geq N+1. \tag{1.77}$$

The sequence $(A_n \circ A_{n-1} \circ \ldots \circ A_1 (\Theta_E))_{n \in \mathbb{N}_0}$ *is asymptotically stable, if and only if for every* $N \in \mathbb{N}_0$ *it is*

$$\lim_{n \to \infty} ||A_n \circ A_{n-1} \circ \ldots \circ A_{N+1}|| = 0. \tag{1.78}$$

Proof. If the sequence $(A_n \circ A_{n-1} \circ \ldots \circ A_1 (\Theta_E))_{n \in \mathbb{N}_0}$ is stable, we choose an arbitrary $\varepsilon > 0$ and conclude that, for every $N \in \mathbb{N}_0$, there exists $\delta = \delta(\varepsilon, N) > 0$ such that for every sequence $(x_n = F_n (x_0))_{n \in \mathbb{N}_0}$, with $x_0 \in E$, it follows that

$$||A_n \circ A_{n-1} \circ \ldots \circ A_{N+1} (x_N) || < \varepsilon \qquad \text{for all} \quad n \geq N+1.$$

This implies

$$||A_n \circ A_{n-1} \circ \ldots \circ A_{N+1} (x_N) || < \varepsilon \qquad \text{for all} \quad n \geq N+1 \quad \text{and} \quad ||x_N|| \leq \frac{\delta}{2},$$

hence,

$$\begin{aligned} &||A_n \circ A_{n-1} \circ \ldots \circ A_{N+1}|| \\ &\quad = \sup_{||x_N|| \leq 1} ||A_n \circ A_{n-1} \circ \ldots \circ A_{N+1} (x_N) || \leq \frac{2\varepsilon}{\delta} c_N \\ &\qquad \text{for all} \quad n \geq N+1 \end{aligned}$$

(here the assumption that all A_n, $n \in \mathbb{N}$, are invertible is needed). Conversely, let (1.77) be true for every $N \in \mathbb{N}_0$. Then we choose $\varepsilon > 0$, $N \in \mathbb{N}$, define $\delta = \varepsilon / c_N$ and conclude that for every sequence $(x_n = F_n (x_0))_{n \in \mathbb{N}_0}$, $x_0 \in E$, with $||x_N|| < \delta$ it follows that

$$\begin{aligned} ||x_n - \Theta_E|| &= ||A_n \circ A_{n-1} \circ \ldots \circ A_{N+1} (x_N) || \\ &\leq ||A_n \circ A_{n-1} \circ \ldots \circ A_{N+1}|| \cdot ||x_N|| \\ &< \varepsilon \qquad \text{for all} \quad n \geq N+1. \end{aligned}$$

Hence the sequence $(x_n = \Theta_E)_{n \in \mathbb{N}_0}$ is stable.

If the sequence $(A_n \circ A_{n-1} \circ \ldots \circ A_1(\Theta_E))_{n\in\mathbb{N}_0}$ is asymptotically stable, hence attractive, then, for every $N \in \mathbb{N}_0$, there exists some $\delta = \delta(N) > 0$ such that for every sequence $(x_n = F_n(x_0))_{n\in\mathbb{N}_0}$, $x_0 \in E$, with $||x_N|| < \delta$ it follows that

$$\lim_{n\to\infty} ||x_n - \Theta_E|| = 0,$$

i.e., for every $\varepsilon > 0$ there exists some $n(\varepsilon) \geq N+1$ with $||x_n - \Theta_E|| < \varepsilon$ for all $n \geq n(\varepsilon)$, hence

$$||A_n \circ A_{n-1} \circ \ldots \circ A_{N+1}(x_N)|| < \varepsilon \qquad \text{for all} \quad n \geq n(\varepsilon) \quad \text{and} \quad ||x_N|| \leq \frac{\delta}{2}.$$

This implies

$$\begin{aligned} &||A_n \circ A_{n-1} \circ \ldots \circ A_{N+1}|| \\ &\quad = \sup_{||x_N||\leq 1} ||A_n \circ A_{n-1} \circ \ldots \circ A_{N+1}(x_N)|| \leq \frac{2\varepsilon}{\delta} \\ &\quad\quad \text{for all} \quad n \geq n(\varepsilon) \end{aligned}$$

(here again the assumption that all A_n, $n \in \mathbb{N}$, are invertible is needed), hence

$$\lim_{n\to\infty} ||A_n \circ A_{n-1} \circ \ldots \circ A_{N+1}|| = 0.$$

Conversely, let (1.78) be true for every $N \in \mathbb{N}_0$. Thus (1.77) is satisfied for every $N \in \mathbb{N}_0$ and therefore the sequence $(A_n \circ A_{n-1} \circ \ldots \circ A_1(\Theta_E))_{n\in\mathbb{N}_0}$ is stable. Further there exists $\delta = \delta(N) > 0$ such that for every sequence $(x_n = F_n(x_0))_{n\in\mathbb{N}_0}$, $x_0 \in E$, with $||x_N|| < \delta$ it follows that

$$\lim_{n\to\infty} ||x_n - \Theta_E|| \leq \lim_{n\to\infty} ||A_n \circ A_{n-1} \circ \ldots \circ A_{N+1}|| \cdot ||x_N|| = 0.$$

Thus, the sequence $(A_n \circ A_{n-1} \circ \ldots \circ A_1(\Theta_E))_{n\in\mathbb{N}_0}$ is attractive. □

Now we specialize to the case $E = \mathbb{R}^k$ equipped with any norm.

Then we have

$$f_n(x) = A_n x + b_n, \qquad x \in \mathbb{R}^k,$$

where $(A_n)_{n\in\mathbb{N}}$ is a sequence of real $k \times k$-matrices and $(b_n)_{n\in\mathbb{N}}$ a sequence of vectors $b_n \in \mathbb{R}^k$. This leads to

$$\begin{aligned} F_n(x) &= f_n \circ f_{n-1} \circ \ldots \circ f_1(x) \\ &= A_n A_{n-1} \cdots A_1 x + \sum_{j=1}^{n} A_n A_{n-1} \cdots A_{j+1} b_j \end{aligned}$$

for $x \in \mathbb{R}^k$ and $n \in \mathbb{N}$ where $A_n A_{n-1} \cdots A_{j+1} b_j = b_n$ for $j = n$.

Let us assume that, for some $p \in \mathbb{N}$,

$$A_{n+p} = A_n \quad \text{and} \quad b_{n+p} = b_n \qquad \text{for all} \quad n \in \mathbb{N}.$$

The question then is whether there is a sequence $(x_n)_{n\in\mathbb{N}_0}$ with

$$x_n = A_n\, x_{n-1} + b_n \qquad \text{for all} \quad n \in \mathbb{N} \tag{1.79}$$

such that

$$x_{n+p} = x_n \qquad \text{for all} \quad n \in \mathbb{N}_0 \tag{1.80}$$

which implies $x_p = x_0$.

Conversely, let this be the case. Then

$$\begin{array}{llllll} x_{1+p} & = & A_{1+p}\, x_p + b_{1+p} & = & A_1\, x_0 + b_1 & = & x_1 \\ x_{2+p} & = & A_{2+p}\, x_{1+p} + b_{2+p} & = & A_2\, x_1 + b_1 & = & x_2 \\ \vdots & & \vdots & & \vdots & & \vdots \\ x_{n+p} & = & A_{n+p}\, x_{n+p-1} + b_{n+p} & = & A_n\, x_{n-1} + b_n & = & x_n, \end{array}$$

hence

$$x_{n+p} = x_n \qquad \text{for all} \quad n \in \mathbb{N}_0.$$

Result. *A sequence $(x_n)_{n\in\mathbb{N}_0}$ with (1.79) satisfies (1.80), if and only if $x_p = x_0$ which is equivalent to $F_p(x_0) = x_0$, i.e.,*

$$x_0 = A_p A_{p-1} \cdots A_1\, x_0 + \sum_{j=1}^{p} A_p A_{p-1} \cdots A_{j+1}\, b_j.$$

This implies that there exists exactly one sequence $(x_n)_{n\in\mathbb{N}_0}$ with (1.79) and (1.80), if and only if the matrix $I - A_p A_{p-1} \cdots A_1$, $I = k \times k-$unit matrix, is non-singular. The first vector x_0 is then given by

$$x_0 = (I - A_p A_{p-1} \cdots A_1)^{-1} \sum_{j=1}^{p} A_p A_{p-1} \cdots A_{j+1} A_j.$$

In the homogeneous case where

$$b_n = \Theta_k \quad \text{for all} \quad n \in \mathbb{N}$$

a non-zero sequence $(x_n)_{n\in\mathbb{N}_0}$ with (1.79) and (1.80) can only exist, if the matrix $I - A_p A_{p-1} \cdots A_1$ is singular, i.e.,

$$\det\,(I - A_p A_{p-1} \cdots A_1) = 0.$$

This again is equivalent to the fact that $\lambda = 1$ is an eigenvalue of $A_p A_{p-1} \cdots A_1$.

1.3.11 Application to a Model for the Process of Hemo-Dialysis

In order to describe the temporal development of the concentration of some poison like urea in the body of a person suffering from a renal disease and having to be attached to an artificial kidney a mathematical model has been proposed in [22] which can be described as follows. The human body is divided into two compartments being termed as cellular part Z and extracellular part E of volume V_Z and V_E, respectively.

The two compartments are separated by cell membranes having the permeability C_Z $[ml/min]$. Let t $[min]$ denote the time and let $K_Z(t)$ $[mg/min]$ and $K_E(t)$ $[mg/min]$ be the concentration of the poison at the time t in Z, and E, respectively. We consider some time interval $[0,T]$, $T>0$, and assume that the patient is attached to the artificial kidney during the subinterval $[0,t_d]$ for some $t_d \in (0,T]$. We further assume that the generation rate of the poison in Z and E is L_1 $[mg/min]$ and L_2 $[mg/min]$, respectively. Then the temporal development of $K_Z = K_Z(t)$ and $K_E = K_E(t)$ in $[0,\infty)$ can be described by the following system of linear differential equations

$$\begin{aligned} V_Z\,\dot{K}_Z(t) &= -C_Z\,(K_Z(t)-K_E(t)) + L_1, \\ V_E\,\dot{K}_E(t) &= -C_Z\,(K_Z(t)-K_E(t)) - C(t)\,K_E(t) + L_2 \end{aligned} \tag{1.81}$$

where

$$C(t) = \begin{cases} C & \text{for} \quad 0 \le t < t_d, \\ 0 & \text{for} \quad t_d \le t < T. \end{cases} \tag{1.82}$$

$$C(t+T) = C(t) \qquad \text{for all} \quad t \ge 0 \tag{1.83}$$

and C $[ml/min]$ is the permeability of the membranes of the artificial kidney.

By (1.83) the periodicity of the process of dialysis is expressed. It can be proved that there is exactly one pair $(K_Z(t), K_E(t))$ of positive and $T-$periodic solutions of (1.81) which intuitively is to be expected. For the numerical computation of these solutions Euler's polygon method can be used. For this purpose we choose a time stepsize $\Delta t > 0$ such that

$$t_d = K\cdot\Delta t, \quad T = N\cdot\Delta t \qquad \text{for all} \quad K,N\in\mathbb{N} \quad \text{with} \quad 2<K<N$$

and replace (1.81) by the difference equations

$$\begin{aligned} V_Z\,(K_Z(t+\Delta t)-K_Z(t)) &= -C_Z\,\Delta t\,(K_Z(t)-K_E(t)) + L_1\,\Delta t, \\ V_E\,(K_E(t+\Delta t)-K_E(t)) &= -C_Z\,\Delta t\,(K_Z(t)-K_E(t)) \\ &\quad -C(t)\,\Delta t\,K_E(t) + L_2\,\Delta t. \end{aligned} \tag{1.84}$$

If we define, for every $n \in \mathbb{N}_0$,

$$x_n = \begin{pmatrix} K_Z(n \cdot \Delta t) \\ K_E(n \cdot \Delta t) \end{pmatrix},$$

$$A_{n+1} = \begin{pmatrix} 1 - C_Z\, \Delta t/V_Z & C_Z\, \Delta t/V_Z \\ C_Z\, \Delta t/V_E & 1 - (C(n \cdot \Delta t) + C_Z)\, \Delta t/V_E \end{pmatrix}$$

and

$$b_{n+1} = \begin{pmatrix} L_1\, \Delta t/V_Z \\ L_2\, \Delta t/V_E \end{pmatrix} = b,$$

then (1.84) can be rewritten in the form

$$x_{n+1} = A_{n+1}\, x_n + b_{n+1}, \qquad n \in \mathbb{N}_0, \tag{1.85}$$

and we have

$$A_{n+N} = A_n \quad \text{and} \quad b_n = b \qquad \text{for all} \quad n \in \mathbb{N}.$$

In order to show the existence and uniqueness of a periodic solution of (1.84) we assume that

$$\Delta t < \min\left(V_Z/C_Z, V_E/(C + C_Z)\right). \tag{1.86}$$

We have to show that the matrix $I - A_N A_{N-1} \cdots A_1$ with $I = \begin{pmatrix} 1 & 0 \\ 0 & 1 \end{pmatrix}$ is non-singular.

Let us put

$$c_1 = C_Z\, \Delta t/V_Z, \qquad c_2 = C_Z\, \Delta t/V_E.$$

Then, for all $n = K, \ldots, N-1$, we obtain

$$A_n = \begin{pmatrix} 1 - c_1 & c_1 \\ c_2 & 1 - c_2 \end{pmatrix}$$

and hence

$$A_n \begin{pmatrix} 1 \\ 1 \end{pmatrix} = \begin{pmatrix} 1 \\ 1 \end{pmatrix}.$$

If we put

$$c_3 = C\, \Delta t/V_E,$$

then we obtain

$$A_n = \begin{pmatrix} 1 - c_1 & c_1 \\ c_2 & 1 - c_2 - c_3 \end{pmatrix} \qquad \text{for} \quad n = 1, \ldots, K-1.$$

This implies

$$A_n \begin{pmatrix} 1 \\ 1 \end{pmatrix} = \begin{pmatrix} 1 \\ 1 - c_3 \end{pmatrix} \leq \begin{pmatrix} 1 \\ 1 \end{pmatrix} \qquad \text{for} \quad n = 1, \ldots, K-1$$

and hence

$$A_{K-2}A_{K-3}\cdots A_1\begin{pmatrix}1\\1\end{pmatrix}\le\begin{pmatrix}1\\1-c_3\end{pmatrix}.$$

Therefore,

$$A_{K-1}A_{K-2}\cdots A_1\begin{pmatrix}1\\1\end{pmatrix}\le\begin{pmatrix}1-c_2c_3\\1-c_3-c_3\,(1-c_2-c_3)\end{pmatrix}<\begin{pmatrix}1\\1\end{pmatrix}$$

and finally

$$A_NA_{N-1}\cdots A_1\begin{pmatrix}1\\1\end{pmatrix}<\begin{pmatrix}1\\1\end{pmatrix}.$$

Since the spectral radius $\varrho(B)$ of every quadratic matrix B with positive elements satisfies the inequality

$$\varrho(B)\le\max_i\frac{(By)_i}{y_i}$$

for all $y=(y)_i$ with $y_i>0$ for all i (see [6]), it follows that

$$\varrho\left(A_NA_{N-1}\cdot A_1\right)<1$$

which implies that $I-A_NA_{N-1}\cdots A_1$ is non-singular.

Hence there exists exactly one $N-$periodic solution of (1.85). This is also positive. In order to see that we have to show that x_0 which is given by

$$x_0=\left(I-A_NA_{N-1}\cdots A_1\right)^{-1}\sum_{j=1}^{N}A_NA_{N-1}\cdots A_{j+1}\,b_j$$

is positive.

First of all we observe that

$$\sum_{j=1}^{N}A_NA_{N-1}\cdots A_{j+1}\,b_j$$

is positive. If we define $B=A_NA_{N-1}\cdots A_1$, then B has positive elements and therefore

$$(I-B)^{-1}=\sum_{j=1}^{\infty}B^j$$

has positive elements which implies that x_0 is positive and in turn that all $x_n,\ n\in\mathbb{N}_0$, are positive.

2

Controlled Systems

2.1 The Time-Continuous Case

2.1.1 The Problem of Controllability

We start with a system of differential equations of the form

$$\dot{x}_i = f_i(x, u), \qquad i = 1, \dots, n \tag{2.1}$$

where $x \in \mathbb{R}^n$, $u \in \mathbb{R}^m$,

$$f_i : \mathbb{R}^n \times \mathbb{R}^m \to \mathbb{R}$$

with $f_i \in \mathcal{C}\left(\mathbb{R}^{n+m}, \mathbb{R}\right)$ and $f_i(\cdot, u) \in \mathcal{C}^1\left(\mathbb{R}^n, \mathbb{R}\right)$ for every $u \in \mathbb{R}^m$ and for $i = 1, \dots, n$.

Assumption: For every function $u \in \mathcal{C}^1\left(\mathbb{R}, \mathbb{R}^m\right)$ and every point $x_0 \in \mathbb{R}^n$ there is exactly one function $x \in \mathcal{C}^1\left(\mathbb{R}, \mathbb{R}^n\right)$ with

$$\dot{x}_i(t) = f_i(x(t), u(t)), \quad t \in \mathbb{R} \tag{2.2}$$

$$\text{for} \quad i = 1, \dots, n \quad \text{and}$$

$$x(0) = x_0. \tag{2.3}$$

We consider every function $u \in \mathcal{C}\left(\mathbb{R}, \mathbb{R}^m\right)$ as a control of the system (2.1) that is used in order to transfer every initial state $x_0 \in \mathbb{R}^n$ within a given time interval $[0, T]$ into a final state $x_T \in \mathbb{R}^n$, i.e., we look for a control function $u \in \mathcal{C}\left(\mathbb{R}, \mathbb{R}^m\right)$ such that the unique solution $x \in \mathcal{C}^1\left(\mathbb{R}, \mathbb{R}^n\right)$ of (2.2), (2.3) satisfies the condition

$$x(T) = x_T. \tag{2.4}$$

In the following we are mainly interested in final states that are rest points of the system (2.1) for $u = \Theta_m$ =null vector of $\mathbb{R}^m$ (which we also call the uncontrolled system).

S. Pickl and W. Krabs, *Dynamical Systems: Stability, Controllability and Chaotic Behavior*,
DOI 10.1007/978-3-642-13722-8_2, © Springer-Verlag Berlin Heidelberg 2010

We therefore assume that the system (2.1) for $u = \Theta_m$ possesses a rest point $\widehat{x} \in \mathbb{R}^n$ which is a solution of the equations

$$f_i(\widehat{x}, \Theta_m) = 0 \qquad \text{for} \quad i = 1, \ldots, n. \tag{2.5}$$

Let Ω be a non-empty subset of $\mathbb{R}^m$ with $\Theta_m \in \Omega$.

Definition.

(1) The system (2.1) is called Ω–controllable on the interval $[0,T]$ with $T > 0$, if for every pair $x_0, x_T \in \mathbb{R}^n$ there is a control $u \in C(\mathbb{R}, \mathbb{R}^m)$ with

$$u(t) \in \Omega \qquad \textit{for all} \quad t \in [0,T]$$

such that the unique solution $x \in C^1(\mathbb{R}, \mathbb{R}^n)$ of (2.2), (2.3) satisfies condition (2.4).

(2) The system (2.1) is called Ω–controllable, if there exists some $T > 0$ such that it is Ω–controllable on the interval $[0,T]$.

(3) The system (2.1) is called completely Ω–controllable, if it is Ω–controllable on every interval $[0,T]$, $T > 0$.

The complete Ω–controllability is obviously the strongest property. In the next section we will present sufficient conditions for linear controlled systems to be completely Ω–controllable.

In the case of non-linear systems we will concentrate on the following problem: For a given $x_0 \in \mathbb{R}^n$ and $x_T = \widehat{x}$ =solution of (2.5) and a time $T > 0$ find a control function $u \in C(\mathbb{R}, \mathbb{R}^m)$ such that the unique solution $x \in C^1(\mathbb{R}, \mathbb{R}^n)$ of (2.2), (2.3) satisfies condition (2.4).

2.1.2 Controllability of Linear Systems

Instead of (2.1) we consider a system of the form

$$\dot{x} = A\,x + B\,u \tag{2.6}$$

where $x \in \mathbb{R}^n$, $u \in \mathbb{R}^m$, A = real $n \times n$–matrix and B = real $n \times m$–matrix. If we put

$$f(x,u) = A\,x + B\,u, \qquad x \in \mathbb{R}^n, \; u \in \mathbb{R}^m,$$

then (2.6) is of the form (2.1) and $f \in C^1\left(\mathbb{R}^{n+m}, \mathbb{R}^n\right)$.

The assumption in Section 2.1.1 is satisfied.

For every $x_0 \in \mathbb{R}^n$ and every function $u \in C(\mathbb{R}, \mathbb{R}^m)$ the unique solution $x \in C^1(\mathbb{R}, \mathbb{R}^n)$ of (2.6) with $x(0) = x_0$ is given by

$$x(t) = \mathrm{e}^{tA}\left(x_0 + \int_0^t \mathrm{e}^{-sA}\,B\,u(s)\;\mathrm{d}s\right), \qquad t \in \mathbb{R}. \tag{2.7}$$

We put

$$Y(t) = \mathrm{e}^{-tA} B \qquad \text{for all} \quad t \in \mathbb{R} \tag{2.8}$$

and consider control functions of the form

$$u(t) = Y^T x, \qquad t \in \mathbb{R}, \tag{2.9}$$

where $x \in \mathbb{R}^n$ is chosen arbitrarily. Insertion into (2.7) gives

$$x(t) = \mathrm{e}^{tA} \left(x_0 + \int_0^t Y(s) \, Y(s)^T x \, \mathrm{d}s \right), \qquad t \in \mathbb{R}.$$

Let, for some given $T > 0$, the $n \times n-$matrix

$$M(T) = \int_0^T Y(t) \, Y(t)^T \, \mathrm{d}t \tag{2.10}$$

(which is symmetric and positive semi-definite) be non-singular.

Then, for every vector $x_T \in \mathbb{R}^n$, there is exactly one $z_T \in \mathbb{R}^n$ such that

$$x_T = \mathrm{e}^{TA} \left(x_0 + M(T) \, z_T \right) \quad \Longleftrightarrow \quad M(T) \, z_T = \mathrm{e}^{-TA} \, x_T - x_0$$

which implies that the system (2.6) is $\mathbb{R}^m-$controllable on $[0, T]$.

Lemma 2.1. *For some $T > 0$ the matrix $M(T)$ (2.10) is non-singular, if and only if the following implication holds true*

$$Y(t)^T x = \Theta_m \qquad \textit{for all} \quad t \in [0, T] \quad \Longrightarrow \quad x = \Theta_n. \tag{2.11}$$

Proof.
(a) Let $M(T)$ be non-singular. If there were some $z \in \mathbb{R}^n$ with $z \neq \Theta_n$ such that $Y(t)^T z = \Theta_n$ for all $t \in [0, T]$, then it would follow that $M(T) \, z = \Theta_n$ which is impossible.

(b) Let the implication (2.11) hold true. If $M(T)$ were singular, then there would exist some $z \in \mathbb{R}^n$, $z \neq \Theta_n$, with $M(T) \, z = \Theta_n$.

This implies

$$z^T M(T) \, z = \int_0^T z^T \, Y(t) \, Y(t)^T \, z \, \mathrm{d}t = 0$$

from which it follows that

$$Y(t)^T z = \Theta_m \qquad \text{for all} \quad t \in [0, T].$$

However, this implies $z = \Theta_n$ contradicting $z \neq \Theta_n$. Therefore, $M(T)$ must be non-singular. □

Theorem 2.2. *The implication (2.11) holds true for every $T > 0$, if and only if the so called Kalman condition*

$$\operatorname{rank} \left(B \mid AB \mid \ldots \mid A^{n-1} B \right) = n \tag{2.12}$$

is satisfied.

Proof.
(a) Let the implication (2.11) be violated for some $T > 0$. Then there is some $z \in \mathbb{R}^n$, $z \neq \Theta_n$ with

$$Y(t)^T z = \Theta_m \qquad \text{for all} \quad t \in [0, T].$$

If one differentiates $z^T Y(t)$ successively by t and puts $t = 0$ then it follows that

$$z^T B = \Theta_m^T,\ z^T AB = \Theta_m^T, \ldots,\ z^T A^{n-1} B = \Theta_m^T \tag{2.13}$$

which contradicts condition (2.12).
(b) Let condition (2.12) be violated. Then there is a vector $z \in \mathbb{R}^n$ with $z \neq \Theta_n$ such that (2.13) holds true. Now let

$$\Phi(-\lambda) = a_0 + a_1 (-\lambda) + \ldots + a_{n-1} (-\lambda)^n$$

be the characteristic polynomial of A. Then it follows, by the theorem of Cayley-Hamilton, that $\Phi(-A) = 0$ and, therefore,

$$A^n = b_0 I + b_1 A + \ldots + b_{n-1} A^{n-1}$$

with suitable coefficients $b_0, \ldots, b_{n-1} \in \mathbb{R}$. This implies

$$A^k = b_0 A^{k-n} + b_1 A^{k+1-n} + \ldots + b_{n-1} A^{k-1}$$

for all $k \geq n$ and on using (2.13) one concludes that

$$z^T A^k B = \Theta_m^T \quad \text{for all} \quad k \geq 0 \qquad \Longrightarrow \qquad z^T Y(t) = 0 \quad \text{for all} \quad t \in \mathbb{R},$$

i.e., the condition (2.11) is violated for all $T > 0$. □

As a consequence of the above considerations we obtain

Theorem 2.3. *If the Kalman condition (2.12) is satisfied, then the system (2.6) is completely $\mathbb{R}^m$-controllable.*

Let us demonstrate this result by an example. We consider a moving linear pendulum as being depicted in the following picture:

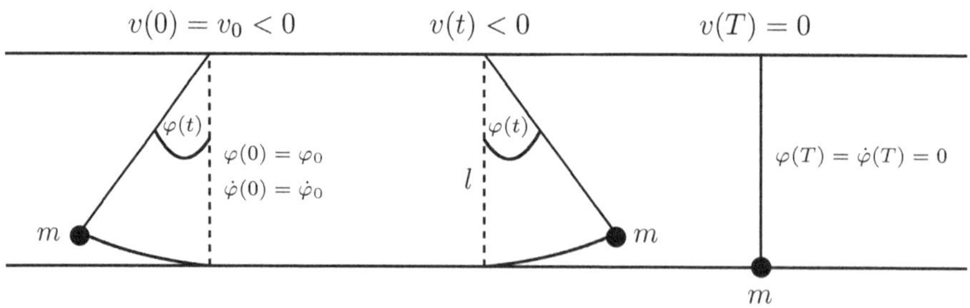

The movement of the pendulum is described by the differential equation

$$\ddot{\varphi}(t) = -\frac{g}{l}\,\varphi(t) - \frac{\ddot{v}(t)}{l}, \qquad t \in \mathbb{R}, \tag{2.14}$$

with the initial conditions

$$\varphi(0) = \varphi_0, \qquad \dot{\varphi}(0) = \dot{\varphi}_0$$

where $\varphi = \varphi(t)$ denotes the deviation angle at the time t and $v = v(t)$ is the location of the suspension point at the time t. This is moving with speed $\dot{v}(t)$ and acceleration $\ddot{v}(t)$ at the time t. The problem to be solved consists of finding a function $v = v(t)$ by which an initial state $(\varphi(0), \dot{\varphi}(0), v(0), \dot{v}(0)) = (\varphi_0, \dot{\varphi}_0, v_0, \dot{v}_0)$ of the pendulum is transferred in the final (rest) state $(\varphi(T), \dot{\varphi}(T), v(T), \dot{v}(T)) = (0, 0, 0, 0)$ at a given time $T > 0$.

If we define a vector function $x(t) = (x_1(t), x_2(t), x_3(t), x_4(t))$ by putting $x_1(t) = \varphi(t)$, $x_2(t) = \dot{\varphi}(t)$, $x_3(t) = v(t)$, $x_4(t) = \dot{v}(t)$, then the differential equation (2.14) can be rewritten in the form

$$\dot{x}(t) = A\,x(t) + B\,u(t), \qquad t \in \mathbb{R}, \tag{2.15}$$

where $u(t) = \ddot{v}(t),\ t \in \mathbb{R}$, is the control and

$$A = \begin{pmatrix} 0 & 1 & 0 & 0 \\ -g/l & 0 & 0 & 0 \\ 0 & 0 & 0 & 1 \\ 0 & 0 & 0 & 0 \end{pmatrix}, \qquad B = \begin{pmatrix} 0 \\ -1/l \\ 0 \\ 1 \end{pmatrix}.$$

In this case we obtain

$$\left(B \mid AB \mid A^2B \mid A^3B\right) = \begin{pmatrix} 0 & -1/l & 0 & g/l^2 \\ -1/l & 0 & g/l^2 & 0 \\ 0 & 1 & 0 & 0 \\ 1 & 0 & 0 & 0 \end{pmatrix},$$

hence

$$\operatorname{rank}\left(B \mid AB \mid A^2B \mid A^3B\right) = 4,$$

i.e., the Kalman condition (2.12) is satisfied.

By Theorem 2.3 the system (2.15) is completely $\mathbb{R}^4$–controllable. This means that every initial state $(\varphi(0), \dot{\varphi}(0), v(0), \dot{v}(0))^T \in \mathbb{R}^4$ of the pendulum can be controlled within every time interval $[0, T]$, $T > 0$, by a suitable control function $u(t) = \ddot{v}(t)$, $t \in [0, T]$ with $u \in C^1(\mathbb{R})$ into every final state $(\varphi(T), \dot{\varphi}(T), v(T), \dot{v}(T))^T \in \mathbb{R}^4$.

2.1.3 Restricted Null-Controllability of Linear Systems

We start with a system

$$\dot{x}(t) = A\,x(t), \qquad t \in \mathbb{R}, \tag{2.16}$$

where $x \in \mathbb{R}^n$ and A =real $n \times n-$matrix.

This system has $\widehat{x} = \Theta_n$ as rest point. In the following we make the special choice $\Omega = [-1, +1]^m$ and ask for the existence of some time $T > 0$ and a control function $u \in \mathcal{C}(\mathbb{R}, \mathbb{R}^m)$ with

$$u(t) \in \Omega \qquad \text{for all} \quad t \in [0, T] \tag{2.17}$$

such that for every initial state $x_0 \in \mathbb{R}^n$ the unique solution $x \in \mathcal{C}^1(\mathbb{R}, \mathbb{R}^n)$ of (2.6) with

$$x(0) = x_0 \tag{2.3}$$

satisfies the end condition

$$x(T) = \widehat{x} = \Theta_n. \tag{2.18}$$

In order to give an affirmative answer to this question we replace the space $\mathcal{C}(\mathbb{R}, \mathbb{R}^m)$ of control functions by the space $L^\infty(\mathbb{R}, \mathbb{R}^m)$ of measurable and essentially bounded $m-$vector functions u on $\mathbb{R}$. Condition (2.17) is replaced by

$$||u(t)||_\infty \leq 1 \qquad \text{for almost all} \quad t \in [0, T] \tag{2.19}$$

where $||\cdot||_\infty$ denotes the maximum norm in $\mathbb{R}^m$.

For every $T > 0$ we define

$$\mathcal{U}_T = \{u \in L^\infty(\mathbb{R}, \mathbb{R}^m) \mid u \text{ satisfies (2.19)}\}. \tag{2.20}$$

From the solution formula (2.7) which also holds true for the unique absolutely continuous solution $x : \mathbb{R} \to \mathbb{R}^n$ of (2.6) with (2.3) in the case that $u \in L^\infty(\mathbb{R}, \mathbb{R}^m)$ it follows that the end condition (2.18) is equivalent to

$$\int_0^T Y(t)\,u(t)\,\mathrm{d}t = -x_0 \tag{2.21}$$

where $Y(t)$ is given by (2.8).

This leads to the
Problem of Restricted Null-Controllability: Let $x_0 \in \mathbb{R}^n$ be given arbitrarily. Find some $T > 0$ and $u \in \mathcal{U}_T$ (2.20) such that condition (2.21) is satisfied.

If this problem has a solution, we call the system (2.6) restricted null-controllable.

If we define, for every $T > 0$, a so called reachable set

$$E(T) = \left\{ x = \int_0^T Y(t)\, u(t)\; \mathrm{d}t \mid u \in \mathcal{U}_T \right\} \subseteq \mathbb{R}^n, \tag{2.22}$$

then the problem of restricted null-controllability is solvable, if and only if

$$E = \bigcup_{T>0} E(T) = \mathbb{R}^n. \tag{2.23}$$

Obviously every reachable set $E(T)$ is convex and we have the following implication

$$0 \le T_1 < T_2 \quad \Longrightarrow \quad E\left(T_1\right) \subseteq E\left(T_2\right).$$

This implies that E is also convex.

Assumption: $E \neq \mathbb{R}^n$.

Then there exists some $\widehat{x} \in \mathbb{R}^n$ with $\widehat{x} \notin E$. By a well known separation theorem for convex sets there exists a number $\alpha \in \mathbb{R}$ and a vector $y \in \mathbb{R}^n$, $y \neq \Theta_n$ such that

$$y^T x \le \alpha < y^T \widehat{x} \qquad \text{for all} \quad x \in E. \tag{2.24}$$

Because of $\Theta_n \in E$ it follows that $\alpha \ge 0$. Further the left side of (2.24) is equivalent to

$$\int_0^T y^T Y(t)\, u(t)\; \mathrm{d}t \le \alpha \qquad \text{for all} \quad u \in \mathcal{U}_T \quad \text{and} \quad T > 0. \tag{2.25}$$

For the following we assume that the Kalman condition (2.12) is satisfied.

Because of $y \neq \Theta_n$ it follows from Theorem 2.2 that for every $T > 0$ there exists some $t_T \in [0, T]$ with

$$y^T Y\left(t_T\right) \neq \Theta_m^T.$$

If we put

$$v(t)^T = y^T Y(t), \qquad t \in \mathbb{R},$$

then there exists at least one component of v which does not vanish identically, say

$$v_1(t) = y^T \mathrm{e}^{-tA} b_1, \qquad t \in \mathbb{R},$$

where b_1 is the first column of the matrix B.

Since with u also $-u$ belongs to $\mathcal{U}_T$, $T > 0$, it follows from (2.25) that

$$\left| \int_0^\infty v(t)^T u(t)\; \mathrm{d}t \right| \le \alpha \qquad \text{for all} \quad u \in L^\infty\left(\mathbb{R}, \mathbb{R}^m\right) \quad \text{with}$$

$$\|u(t)\|_\infty \le 1 \qquad \text{for almost all} \quad t \in (0, \infty).$$

This implies

$$\int_0^\infty ||v(t)||_1 \, \mathrm{d}t = \sum_{i=1}^n \int_0^\infty |v_i(t)| \, \mathrm{d}t \le \alpha < \infty$$

and further

$$\int_0^\infty |v_1(t)| \, \mathrm{d}t \le \alpha.$$

This implies the existence of the integrals

$$w(t) = \int_t^\infty v_1(s) \, \mathrm{d}s \qquad \text{for all} \quad t \in [0, \infty)$$

and it follows that

$$\frac{\mathrm{d}}{\mathrm{d}t} w(t) = -v_1(t) \qquad \text{for all} \quad t \in [0, \infty) \quad \text{as well as} \quad \lim_{t \to \infty} w(t) = 0.$$

If we put $D = \dfrac{\mathrm{d}}{\mathrm{d}t}$, it follows that

$$\left(D^k v_1\right)(t) = y^T(-A)^k \, \mathrm{e}^{-tA} \, b_1, \qquad t \in \mathbb{R}, \quad \text{for} \quad k = 0, 1, 2, \ldots$$

If $\psi(\lambda)$ denotes the characteristic polynomial of A, then the application of the Cayley-Hamilton-theorem implies $\psi(A) = 0$ and therefore

$$\left(\psi(-D) \, v_1\right)(t) = y^T(\psi(A)) \, \mathrm{e}^{-tA} \, b_1 = 0 \qquad \text{for all} \quad t \in [0, \infty).$$

This implies

$$\psi(-D)(-D \, w)(t) = 0 \qquad \text{for all} \quad t \in [0, \infty)$$

and because $(-D)$ and $\psi(-D)$ are interchangeable we can conclude

$$(-D \, \psi(-D) \, w)(t) = 0 \qquad \text{for all} \quad t \in [0, \infty).$$

The characteristic equation of this linear differential equation reads

$$-\lambda \, \psi(-\lambda) = 0.$$

From $\lim_{t \to \infty} w(t) = 0$ we infer that at least one solution $-\lambda$ of the equation $\psi(-\lambda) = 0$ must have a negative real part. Therefore the matrix A must have an eigenvalue with a positive real part.

Summarizing we obtain by contraposition the

Theorem 2.4. *If the Kalman condition (2.12) is satisfied and if the matrix A has only eigenvalues with non-positive real parts, then the system (2.6) is restricted null-controllable. This means that for every initial state $x_0 \in \mathbb{R}^n$ there is some $T > 0$ and a control function $u \in \mathcal{U}_T$ (2.20) such that the unique absolutely continuous solution $x = x(t)$, $t \in \mathbb{R}$, of (2.6), (2.3) satisfies the end condition (2.18).*

The assumptions of this theorem are for instance satisfied for the system (2.15). It has been shown already that the Kalman condition (2.12) is satisfied. It is easy to show that the eigenvalues of the matrix A are given by

$$\lambda_1 = i\sqrt{\frac{g}{l}}, \quad \lambda_2 = -i\sqrt{\frac{g}{l}}, \quad \lambda_3 = \lambda_4 = 0.$$

2.1.4 Controllability of Nonlinear Systems into Rest Points

We again start with a system of the form

$$\dot{x} = f(x, u) \tag{2.1}$$

where $f \in \mathcal{C}(\mathrm{I\!R}^n \times \mathrm{I\!R}^m, \mathrm{I\!R}^n)$ and $f(\cdot, u) \in \mathcal{C}^1(\mathrm{I\!R}^n, \mathrm{I\!R}^m)$ for every $u \in \mathrm{I\!R}^m$. We make the same assumptions as at the beginning of Section 2.1.1.

Let $\widehat{x} \in \mathrm{I\!R}^n$ be a rest point of the uncontrolled system for $u \equiv \Theta_m$, i.e., a solution of the equation

$$f(\widehat{x}, \Theta_m) = \Theta_n. \tag{2.5}$$

Given an arbitrary initial state $x_0 \in \mathrm{I\!R}^n$ we then look for some time $T > 0$ and a control function $u \in \mathcal{C}(\mathrm{I\!R}, \mathrm{I\!R}^m)$ such that the unique solution $x \in \mathcal{C}^1(\mathrm{I\!R}, \mathrm{I\!R}^n)$ of (2.1) with

$$x(0) = x_0 \tag{2.3}$$

satisfies the end condition

$$x(T) = \widehat{x}.$$

If this problem has a solution for every $x_0 \in \mathrm{I\!R}^n$ we call the system (2.1) controllable to $\widehat{x}$.

Let us weaken this concept to a concept of local controllability. For this purpose we assume that

$$f_i \in \mathcal{C}^1(\mathrm{I\!R}^n \times \mathrm{I\!R}^m) \qquad \text{for all} \quad i = 1, 2, \ldots, n$$

and linearize the equation (2.1) in $(\widehat{x}, \Theta_m)$, i.e., we replace it by

$$\dot{x} = Ax + Bu \tag{2.6}$$

where

$$A = \left(f_{ix_j}(\widehat{x}, \Theta_m)\right)_{i,j=1,\ldots,n} \qquad \text{and} \qquad B = \left(f_{iu_k}(\widehat{x}, \Theta_m)\right)_{\substack{i=1,\ldots,n \\ k=1,\ldots,m}}. \tag{2.26}$$

Definition. *The system (2.1) is called locally controllable to $\widehat{x}$, if the system (2.6) with A and B by (2.26) is $\mathrm{I\!R}^m$−null-controllable, i.e., if for every $x_0 \in \mathrm{I\!R}^n$ there exists a $T > 0$ and a control function $u \in \mathcal{C}(\mathrm{I\!R}, \mathrm{I\!R}^m)$ such that the unique solution $x \in \mathcal{C}^1(\mathrm{I\!R}, \mathrm{I\!R}^n)$ of (2.6), (2.3) satisfies the end condition*

$$x(T) = \Theta_m. \tag{2.18}$$

An immediate consequence of Theorem 2.3 is the

Theorem 2.5. *If A and B given by (2.26) satisfy the Kalman condition (2.12), then the system (2.1) is locally controllable to $\widehat{x}$.*

Definition. *The system (2.1) is called locally restricted controllable to $\widehat{x}$, if the system (2.6) with A and B by (2.26) is restricted null-controllable, i.e., if for every $x_0 \in \mathbb{R}^n$ there exists a $T > 0$ and a control function $u \in \mathcal{U}_T$ (2.20) such that the unique absolutely continuous solution $x = x(t)$, $t \in \mathbb{R}$, of (2.6), (2.3) satisfies the end condition (2.18).*

An immediate consequence of Theorem 2.4 is the

Theorem 2.6. *If A and B given by (2.26) satisfy the Kalman condition (2.12) and if the matrix A has only eigenvalues with non-positive real parts, then the system (2.1) is locally restricted controllable to $\widehat{x}$.*

Let us demonstrate these two theorems by the example of a nonlinear pendulum with movable suspension point. We use the same notations as in the case of a linear pendulum in Section 2.1.2.

The movement of the pendulum is described by the differential equation

$$\ddot{\varphi}(t) = \frac{g}{l}\sin\varphi(t) - \frac{\ddot{v}(t)}{l}\cos\varphi(t), \qquad t \in \mathbb{R}. \tag{2.27}$$

If one defines functions

$$x_1(t) = \varphi(t), \quad x_2(t) = \dot{\varphi}(t), \quad x_3(t) = v(t), \quad x_4(t) = \dot{v}(t),$$

then (2.27) can be rewritten in the form

$$\begin{aligned}
\dot{x}_1(t) &= x_2(t),\\
\dot{x}_2(t) &= -\frac{g}{l}\sin x_1(t) - \frac{u(t)}{l}\cos x_1(t),\\
\dot{x}_3(t) &= x_4(t),\\
\dot{x}_4(t) &= u(t)
\end{aligned} \tag{2.28}$$

where $u(t) = \ddot{v}(t)$, $t \in \mathbb{R}$, is the control function. If we define

$$f(x,u) = \left(f_1(x,u), f_2(x,u), f_3(x,u), f_4(x,u)\right)^T, \qquad x \in \mathbb{R}^4,\ u \in \mathbb{R},$$

where

$$\begin{aligned}
f_1(x,u) &= x_2(t),\\
f_2(x,u) &= -\frac{g}{l}\sin x_1 - \frac{u}{l}\cos x_1,\\
f_3(x,u) &= x_4,\\
f_4(x,u) &= u,
\end{aligned}$$

then the system (2.28) takes the form (2.1) and we have

$$f_i \in C^1\left(\mathbb{R}^4, \mathbb{R}\right) \qquad \text{for} \quad i = 1,2,3,4.$$

Obviously for $u \equiv 0$ the point $\Theta_4 = (0,0,0,0)^T \in \mathbb{R}^4$ is a rest point of the system (2.28) and we obtain

$$A = \left(f_{ix_j}\left(\Theta_4, 0\right)\right)_{i,j=1,2,3,4} = \begin{pmatrix} 0 & 1 & 0 & 0 \\ -g/l & 0 & 0 & 0 \\ 0 & 0 & 0 & 1 \\ 0 & 0 & 0 & 0 \end{pmatrix},$$

$$B = \left(f_{iu}\left(\Theta_4, 0\right)\right)_{i=1,2,3,4} = \begin{pmatrix} 0 \\ -1/l \\ 0 \\ 1 \end{pmatrix}.$$

We have already shown (see Section 2.1.2) that the Kalman condition (2.12) is satisfied for A and B and that the matrix A has only eigenvalues with real parts equal to zero (see Section 2.1.3).

Theorem 2.6 then implies that the system is locally restricted controllable to $\widehat{x} = \Theta_4$.

Now we return to the question of the $\widehat{x}-$controllability of the system (2.1). We again assume a set $\Omega \subseteq \mathbb{R}^n$ with $\Theta_m \in \Omega$ to be given and define, for every $T > 0$,

$$\mathcal{U}_T = \left\{u \in L^\infty\left(\mathbb{R}, \mathbb{R}^m\right) \mid u(t) \in \Omega \text{ for almost all } t \in [0,T]\right\}. \tag{2.29}$$

Further let $S(\widehat{x}, T)$, for every $T > 0$, be the set of all $x_0 \in \mathbb{R}^n$ such that there exists a $u \in \mathcal{U}_T$ such that the unique absolutely continuous solution $x = x(t)$, $t \in \mathbb{R}$, of (2.1), (2.3) satisfies the end condition $x(T) = \widehat{x}$.

If we then define

$$S(\widehat{x}) = \bigcup_{T>0} S(\widehat{x}, T), \tag{2.30}$$

the set $S(\widehat{x})$ consists of all vectors $x_0 \in \mathbb{R}^n$ such that there exists a time $T > 0$ and a control $u \in \mathcal{U}_T$ such that the unique absolutely continuous solution $x = x(t)$, $t \in \mathbb{R}$, of (2.1), (2.3) satisfies the end condition $x(T) = \widehat{x}$.

With this definition we formulate the

Theorem 2.7. *Let $\widehat{x}$ be an interior point of $S(\widehat{x})$ ($\widehat{x} \in S(\widehat{x})$ follows from the definition of $S(\widehat{x})$) and let $\widehat{x}$ be attractive, i.e., there exists an open set $\mathcal{U} \subseteq \mathbb{R}^n$ with $\widehat{x} \in \mathcal{U}$ such that for every $x_0 \in \mathcal{U}$ the unique solution $x \in C^1(\mathbb{R}, \mathbb{R}^n)$ of $\dot{x} = f(x, \Theta_m)$ with $x(0) = x_0$ satisfies the statement $\lim_{t\to\infty} x(t) = \widehat{x}$. Then it follows that $S(\widehat{x}) \supseteq \mathcal{U}$.*

Proof. Let $x_0 \in \mathcal{U}$ be chosen arbitrarily. Then for the solution $x \in C^1(\mathbb{R}, \mathbb{R}^n)$ of $\dot{x} = f(x, \Theta_m)$ with $x(0) = x_0$ it follows that $\lim_{t\to\infty} x(t) = \widehat{x}$. Since $\widehat{x}$ is an interior point of $S(\widehat{x})$, there exists some $t_1 > 0$ with $x(t_1) \in S(\widehat{x})$. This implies the existence of some time $T > 0$ and a function $u \in \mathcal{U}_T$ such that the absolutely continuous solution $\widetilde{x} = \widetilde{x}(t)$ of

$$\dot{\widetilde{x}}(t) = f(\widetilde{x}(t), u(t)), \qquad t \in \mathbb{R} \quad \text{with} \quad \widetilde{x}(0) = x(t_1)$$

satisfies the end condition $\widetilde{x}(T) = \widehat{x}$.

If we define

$$u^*(t) = \begin{cases} \Theta_m & \text{for} \quad t \leq t_1, \\ u(t - t_1) & \text{for} \quad t > t_1, \end{cases}$$

then $u^* \in \mathcal{U}_T$ and it follows for the absolutely continuous solution $x^* = x^*(t)$, $t \in \mathbb{R}$, of $\dot{x}^*(t) = f(x^*(t), u^*(t))$, $t \in \mathbb{R}$, $x^*(0) = x(0) = x_0$ that the end condition $x^*(T + t_1) = \widetilde{x}(T) = \widehat{x}$ is satisfied which completes the proof. □

Let us apply this result to linear systems of the form (2.6). In this case we have

$$f(x, u) = A\,x + B\,u, \qquad x \in \mathbb{R}^n, \; u \in \mathbb{R}^m.$$

A rest point of the uncontrolled system

$$\dot{x} = f(x, \Theta_m) = A\,x, \qquad x \in \mathbb{R}^n, \tag{2.16}$$

is given by $\widehat{x} = \Theta_n$. This is the only one, if the matrix A is non-singular. If all the eigenvalues of A have negative real parts (which implies that A is non-singular), then $\widehat{x} = \Theta_n$ is globally attractive, i.e., for every $x_0 \in \mathbb{R}^n$ the unique solution $x \in C^1(\mathbb{R}, \mathbb{R}^n)$ of (2.16) with $x(0) = x_0$ satisfies the condition $\lim_{t\to\infty} x(t) = \Theta_n$.

Theorem 2.8. *If the Kalman condition (2.12) is satisfied, then $\widehat{x} = \Theta_n$ is an interior point of $S(\Theta_n) = E$ (2.23).*

Proof. Assume that Θ_n is no interior point of E. Then Θ_n is also not an algebraically interior point of E and $K(E) = \bigcup_{\lambda \geq 0} \lambda E$ is not equal to $\mathbb{R}^n$. Since because of $E = -E$ the convex cone $K(E)$ is a linear space, the set E, as subset of $K(E)$, is contained in a hyperplane, i.e., there exists some $y \in \mathbb{R}^n$ with $y \neq \Theta_n$ such that

$$y^T e = 0 \qquad \text{for all} \quad e \in E.$$

In particular it follows for every $T > 0$ that

$$\int_0^T y^T e^{tA} B\,u(t)\,dt = 0 \qquad \text{for all} \quad u \in \mathcal{U}_T$$

and hence for all $u \in L^\infty([0,T], \mathbb{R}^m) = \overline{\bigcup_{\lambda \geq 0} \lambda \mathcal{U}_T}$.

This implies

$$y^T e^{-tA} B = 0 \qquad \text{for all} \quad t \in [0,T]$$

and in turn $y = \Theta_n$ by Theorem 2.2 which contradicts $y \neq \Theta_n$. □

As a consequence of Theorem 2.7 and Theorem 2.8 we obtain

Theorem 2.9. *If the Kalman condition (2.12) is satisfied and if the matrix A has only eigenvalues with negative real parts, then for every $x_0 \in \mathbb{R}^n$ there exists some $T > 0$ and a control function $u \in \mathcal{U}_T$ (2.20) such that the unique absolutely continuous solution $x = x(t)$, $t \in \mathbb{R}$, of (2.6), (2.3) (which is given by (2.7)) satisfies the end condition $x(T) = \widehat{x}$ (2.18).*

This theorem is contained in Theorem 2.4.

2.1.5 Approximate Solution of the Problem of Restricted Null-Controllability

We again consider the problem of restricted null-controllability as in Section 2.1.3. We assume $T > 0$ and $x_0 \in \mathbb{R}^n$ to be given and formulate the following

Approximation Problem: Find some $u_T \in \mathcal{U}_T$ (2.20) such that

$$\left\| \int_0^T Y(t)\, u_T(t)\, dt + x_0 \right\|_2 \leq \left\| \int_0^T Y(t)\, u(t)\, dt + x_0 \right\|_2 \qquad \text{for all} \quad u \in \mathcal{U}_T$$

where $Y(t)$ is given by (2.8).

A solution $u_T \in \mathcal{U}_T$ of this approximation problem is considered as an approximation of a solution of the problem of restricted null-controllability.

If one defines

$$V = \left\{ y \in \mathbb{R}^n \mid y = \int_0^T Y(t)\, u(t)\, dt \text{ with } u \in \mathcal{U}_T \right\},$$

one obtains a convex subset of $\mathbb{R}^n$. This set is weakly closed, i.e., the following implication holds true:

$$\left\{ \begin{array}{l} \lim\limits_{k\to\infty} y_k^T x = y^T x \;\; \text{for all} \;\; x \in \mathbb{R}^n \\ \quad\text{and} \;\; y_k \in V \;\; \text{for all} \;\; k \in \mathbb{N} \end{array} \right\} \implies y \in V.$$

Proof. $y_k \in V$ implies for every $k \in \mathbb{N}$ the existence of some function

$$u_k \in \{u \in L^\infty([0,T], \mathbb{R}^m) \mid ||u(t)||_\infty \leq 1 \text{ for almost all } t \in [0,T]\} = K_\infty(T)$$

with

$$y_k = \int_0^T Y(t)\, u_k(t)\; \mathrm{d}t$$

which implies

$$\lim_{k\to\infty} \int_0^T u_k(t)^T\, Y(t)\, x\; \mathrm{d}t = y^T x \qquad \text{for all} \quad x \in \mathbb{R}^n.$$

Since the set $K_\infty(T)$ is weak-$*$ sequentially compact, there is some $u \in K_\infty(T)$ and a subsequence $(u_{k_i})_{i\in\mathbb{N}}$ with

$$\lim_{i\to\infty} \int_0^T u_{k_i}(t)^T\, Y(t)\, x\; \mathrm{d}t = \int_0^T u(t)^T\, Y(t)\, x\; \mathrm{d}t$$

which implies

$$\left(y^T - \int_0^T u(t)^T\, Y(t)^T\; \mathrm{d}t \right) x = 0 \qquad \text{for all} \quad x \in \mathbb{R}^n.$$

From this it follows that

$$y = \int_0^T Y(t)\, u(t)\; \mathrm{d}t \quad \Longleftrightarrow \quad y \in V$$

which completes the proof. □

The weak closedness of the set V implies that V is also closed.

As a consequence of a well known theorem in approximation theory we then obtain the existence of exactly one $y^* \in V$ with

$$||y^* + x_0||_2 \leq ||y + x_0||_2 \qquad \text{for all} \quad y \in V$$

which is characterized by

$$(y^* + x_0)^T\; y^* \leq (y^* + x_0)^T\; y \qquad \text{for all} \quad y \in V,$$

if $y^* + x_0 \neq \Theta_n$.

Now we distinguish two cases:

(a) $y^* + x_0 = \Theta_n$. Then every $u^* \in \mathcal{U}_T$ with

$$y^* = \int_0^T Y(t)^T\, u^*(t)\; \mathrm{d}t$$

is a solution of the problem of restricted null-controllability.

b) $y^* + x_0 \neq \Theta_n$. Then for every $u^* \in \mathcal{U}_T$ as in (a) we obtain from the above characterization of $y^* \in V$ the equation

$$-\int_0^T (y^* + x_0)^T\, Y(t)\, u^*(t)\ \mathrm{d}t = \int_0^T \|(y^* + x_0)^T\, Y(t)\|_1\ \mathrm{d}t.$$

We try to solve this equation iteratively by starting with

$$y^0 = \Theta_n \quad \text{and} \quad u_k^0(t) = -\mathrm{sgn}\left(x_0^T\, Y(t)\right)_k \qquad \text{for} \quad k = 1, \ldots, n,\ t \in [0, T],$$

(where $\mathrm{sgn}(0) = 0$) and constructing a sequence $\left(y^N\right)_{N \in \mathbb{N}_0}$ in $\mathbb{R}^n$ and a sequence $\left(u^N\right)_{N \in \mathbb{N}_0}$ in $K_\infty(T)$ as follows:
If $y^N \in \mathbb{R}^n$ is given, then we define

$$u_k^N(t) = -\mathrm{sgn}\left(\left(y^N + x_0\right)^T\, Y(t)\right)_k \qquad \text{for} \quad k = 1, \ldots, n,\ t \in [0, T],$$

and

$$y^{N+1} = \int_0^T Y(t)\, u^N(t)\ \mathrm{d}t.$$

If the sequence $\left(u^N\right)_{N \in \mathbb{N}_0}$ converges weak-$*$ to $u^* \in K_\infty(T)$, then the sequence $\left(y^N\right)_{N \in \mathbb{N}_0}$ to $y^* = \int_0^T Y(t)\, u^*(t)\ \mathrm{d}t$ and y^* and u^* satisfy the equation in (b).

2.1.6 Time-Minimal Restricted Null-Controllability of Linear Systems

We replace the set $\mathcal{U}_T$ (2.20) in Section 2.1.3 by

$$\mathcal{U} = \left\{u \in L^\infty(\mathbb{R}, \mathbb{R}^m) \mid \operatorname*{ess\,sup}_{t \in \mathbb{R}} \|u(t)\|_2 \leq \gamma\right\} \tag{2.31}$$

where $\|u(t)\|_2 = \left(u(t)^T u(t)\right)^{1/2}$, $t \in \mathbb{R}$.

The problem of restricted null-controllability now consists of finding a time $T > 0$ and some $u \in \mathcal{U}$ (2.31) such that

$$\int_0^T Y(t)\, u(t)\ \mathrm{d}t = -x_0 \tag{2.21}$$

where $x_0 \in \mathbb{R}^n$ is given arbitrarily.

Let us assume that the problem is solvable. Then we define a so called minimum time $T(\gamma)$ by

$$T(\gamma) = \inf\{T > 0 \mid \text{ There is some } u \in \mathcal{U} \text{ which satisfies (2.21)}\}. \tag{2.32}$$

At first we prove

Theorem 2.10. *If restricted null-controllability is possible, then time-minimal restricted null-controllability is also possible, i.e., there exists some* $u_\gamma \in \mathcal{U}$ *such that*

$$\int_0^{T(\gamma)} Y(t)\, u_\gamma(t)\ dt = -x_0. \tag{2.33}$$

Proof. By the definition of $T(\gamma)$ there exists a sequence $(T_k)_{k\in\mathbb{N}}$ of times $T_k \geq T(\gamma)$ with $\lim\limits_{k\to\infty} T_k = T(\gamma)$ and corresponding controls $u_k \in \mathcal{U}$ with

$$\int_0^{T_k} Y(t)\, u_k(t)\ dt = -x_0 \qquad \text{for all} \quad k \in \mathbb{N}.$$

Since the set $\mathcal{U}$ is sequentially weak-$*$ compact, there is a subsequence $(u_{k_i})_{i\in\mathbb{N}}$ of the sequence $(u_k)_{k\in\mathbb{N}}$ and some $u_\gamma \in \mathcal{U}$ such that

$$\lim_{i\to\infty} \int_0^{T(\gamma)} y(t)^T\, u_{k_i}(t)\ dt = \int_0^{T(\gamma)} y(t)^T\, u_\gamma(t)\ dt$$
$$\text{for all} \quad y \in L^1\left([0, T(\gamma)], \mathbb{R}^m\right)$$

which implies

$$\lim_{i\to\infty} \int_0^{T(\gamma)} Y(t)^T\, u_{k_i}(t)\ dt = \int_0^{T(\gamma)} Y(t)\, u_\gamma(t)\ dt.$$

For every $i \in \mathbb{N}$ we then have

$$\int_0^{T(\gamma)} Y(t)\, u_\gamma(t)\ dt$$
$$= \underbrace{\int_0^{T(\gamma)} Y(t)\left(u_\gamma(t) - u_{k_i}(t)\right)\ dt}_{\to\, \Theta_m} + \underbrace{\int_0^{T_{k_i}} Y(t)\, u_{k_i}(t)\ dt}_{=\,-x_0} - \int_{T(\gamma)}^{T_{k_i}} Y(t)\, u_{k_i}(t)\ dt$$

From this we infer because of $\lim\limits_{i\to\infty} T_{k_i} = T(\gamma)$ and therefore

$$\lim_{i\to\infty} \int_{T(\gamma)}^{T_{k_i}} Y(t)\, u_{k_i}(t)\ dt = \Theta_m$$

that (2.33) is satisfied which completes the proof. □

Every $u_\gamma \in \mathcal{U}$ which satisfies (2.33) is called a time-minimal null-control. The next step is to characterize time-minimal null-controls.

For this purpose we make use of the

Normality Condition: For every $T > 0$ and every $y \in \mathbb{R}^n$ the components of the vector function $Y(t)^T y$ only vanish on a subset of $[0, T]$ of (Lebesgue-) measure zero unless $y = \Theta_n$.

At first we consider the

Minimum Norm Problem: For $T > 0$ and $x_0 \in \mathbb{R}^n$ given determine $u \in L^\infty(\mathbb{R}, \mathbb{R}^m)$ such that (2.21) is satisfied and the functional φ defined by

$$\varphi(u) = ||u||_{L^\infty([0,T],\mathbb{R}^m)} = \operatorname*{ess\ sup}_{t\in[0,T]} ||u(t)||_2 \tag{2.34}$$

is minimized.

In order to study the problem of minimizing $\varphi(u)$ defined by (2.34) on the linear manifold

$$M = \left\{ u \in L^\infty(\mathbb{R}, \mathbb{R}^m) \,\middle|\, \int_0^T Y(t)\, u(t)\ dt = -x_0 \right\} \tag{2.35}$$

we also consider the
Dual Problem: Determine $y \in \mathbb{R}^n$ such that

$$\int_0^T \left(y^T Y(t) Y(t)^T y\right)^{1/2}\ dt \leq 1 \tag{2.36}$$

holds true and the functional

$$\psi(y) = -x_0^T y \tag{2.37}$$

is minimized.

Let

$$N = \{y \in \mathbb{R}^n \mid \text{(2.36) is satisfied}\}. \tag{2.38}$$

If $u \in M$ and $y \in N$, then it follows that

$$\begin{aligned}
\psi(y) = -x_0^T y &= \int_0^T u(t)^T Y(t)^T y\ dt \\
&\leq \int_0^T \left(u(t)^T u(t)\right)^{1/2} \left(y^T Y(t) Y(t)^T y\right)^{1/2}\ dt \\
&\leq \operatorname*{ess\ sup}_{t\in[0,T]} ||u(t)||_2 \int_0^T \left(y^T Y(t) Y(t)^T y\right)^{1/2} \leq \varphi(u)
\end{aligned}$$

which implies

$$\sup_{y\in N} \psi(y) \leq \inf_{u\in M} \varphi(u). \tag{2.39}$$

The dual problem will now be used to solve the minimum norm problem.

At first we observe that the normality condition implies that for every $T > 0$ the function

$$y \to \chi(y) = \int_0^T \left(y^T Y(t) Y(t)^T y\right)^{1/2} \, dt, \qquad y \in \mathbb{R}^n,$$

is Gateaux-differentiable and its Gateaux-derivative is given by

$$D\chi(y,h) = \nabla \chi(y)^T h, \qquad h \in \mathbb{R}^n,$$

where

$$\nabla \chi(y) = \int_0^T \frac{1}{\left(y^T Y(t) Y(t)^T y\right)^{1/2}} Y(t) Y(t)^T y \, dt.$$

Now we can prove

Theorem 2.11. *Under the normality condition the minimum norm problem has exactly one solution (on $[0,T]$) for every choice of $T > 0$ which is of the form*

$$u_T(t) = \frac{-x_0^T y_T}{\left(y_T^T Y(t) Y(t)^T y_T\right)^{1/2}} Y(t)^T y_T \tag{2.40}$$

for almost all $t \in [0,T]$ and $u_T(t) = \Theta_m$ for all $t \notin [0,T]$

where $y_T \in \mathbb{R}^n$ is a solution of the dual problem and satisfies

$$\int_0^T \left(y_T^T Y(t) Y(t)^T y_T\right)^{1/2} dt = 1. \tag{2.41}$$

Proof. Since the set of all $y \in \mathbb{R}^n$ which satisfy (2.36) is compact and the function $\psi(y) = -x_0^T y$ is continuous there is some $y_T \in \mathbb{R}^n$ which solves the dual problem and every solution satisfies (2.41). Hence the dual problem has the same solutions, if we assume equality to hold in (2.36).

Let $y_T \in \mathbb{R}^n$ be a solution of the dual problem. Then by the Lagrangean multiplier rule there is a multiplier $\lambda_T \in \mathbb{R}$ such that

$$\nabla \psi\left(y_T\right) - \lambda_T \nabla \chi\left(y_T\right) = \Theta_n$$

which is equivalent to

$$-x_0 = \lambda_T \int_0^T \frac{1}{\left(y_T^T Y(t) Y(t)^T y_T\right)^{1/2}} Y(t) Y(t)^T y_T \, dt$$

and implies

$$-x_0^T y_T = \lambda_T \int_0^T \left(y_T^T Y(t) Y(t)^T y_T\right)^{1/2} dt = \lambda_T.$$

Therefore, if we define $u_T = u_T(t)$ by (2.40), then (2.21) is satisfied for $u = u_T$. Further it follows that

$$\left(u_T(t)^T u_T(t)\right)^{1/2} = -x_0^T\, y_T = \sup_{y \in N} \psi(y) \qquad \text{for almost all} \quad t \in [0,T]$$

which implies $\varphi(u_T) = \sup\limits_{y \in N} \psi(y)$. From (2.39) it then follows that u_T solves the minimum norm problem.

It remains to prove the uniqueness of u_T. Let $\widehat{u} \in M$ be any solution of the minimum norm problem. In order to find a solution $u \in L^\infty(\mathbb{R}, \mathbb{R}^m)$ of (2.21) we tentatively put $u(t) = Y(t)^T y$ for $t \in \mathbb{R}$. Then (2.21) reads

$$\int_0^T Y(t)Y(t)^T\, \mathrm{d}t\, y = -x_0. \tag{2.42}$$

The normality condition implies that the row vector functions $y_i^T = y_i(t)^T$, $t \in \mathbb{R}$, $i = 1, \ldots, n$ of $Y = Y(t)$ are linearly independent. From this it follows that the matrix $\int_0^T Y(t)Y(t)^T\, \mathrm{d}t$ is non-singular so that for every choice of $x_0 \in \mathbb{R}^n$ the linear system (2.42) has exactly one solution $y \in \mathbb{R}^n$. Therefore the mapping

$$u \mapsto \int_0^T Y(t)\, u(t)\, \mathrm{d}t \qquad \text{from} \quad L^\infty(\mathbb{R}, \mathbb{R}^m) \quad \text{into} \quad \mathbb{R}^n$$

is surjective. This implies that there exists a multiplier $y \in \mathbb{R}^n$ such that

$$\varphi(\widehat{u}) \leq \varphi(u) - y^T \left(\int_0^T Y(t)\, u(t)\, \mathrm{d}t + x_0 \right) \tag{2.43}$$

for all $u \in L^\infty(\mathbb{R}, \mathbb{R}^m)$, hence

$$\varphi(\widehat{u}) + y^T x_0 \leq \varphi(u) - \int_0^T y^T\, Y(t)\, u(t)\, \mathrm{d}t$$

for all $u \in L^\infty(\mathbb{R}, \mathbb{R}^m)$ which is only possible, if

$$\varphi(u) - \int_0^T y^T\, Y(t)\, u(t)\, \mathrm{d}t \geq 0 \qquad \text{for all} \quad u \in L^\infty(\mathbb{R}, \mathbb{R}^m).$$

This is equivalent to

$$\int_0^T y^T Y(t)\, u(t)\, \mathrm{d}t \le \operatorname*{ess\ sup}_{t\in[0,T]} \left(u(t)^T u(t)\right)^{1/2} \quad \text{for all} \quad u \in L^\infty(\mathbb{R}, \mathbb{R}^m)$$

or

$$\int_0^T y^T Y(t)\, u(t)\, \mathrm{d}t \le 1 \text{ for all } u \in L^\infty(\mathbb{R}, \mathbb{R}^m) \text{ with } \operatorname*{ess\ sup}_{t\in[0,T]} \left(u(t)^T u(t)\right)^{1/2} = 1.$$

If we define

$$u(t) = \frac{1}{\left(y^T Y(t) Y(t)^T y\right)^{1/2}}\, Y(t)^T y$$
$$\text{for almost all } t \in [0,T] \text{ and } u(t) = \Theta_m \text{ for all } t \notin [0,T]$$

then it follows that

$$\operatorname*{ess\ sup}_{t\in[0,T]} \left(u(t)^T u(t)\right)^{1/2} = 1$$

and therefore

$$\int_0^T \left(y^T Y(t) Y(t)^T y\right)^{1/2} \mathrm{d}t = \int_0^T y^T Y(t)\, u(t)\, \mathrm{d}t \le 1.$$

Thus $y \in N$. If we choose $u \equiv \Theta_m$ in (2.43) it follows that $\varphi(\widehat{u}) \le -x_0^T y$. On the other hand we have shown (in the proof of (2.39)) that $-x_0^T y \le \varphi(\widehat{u})$. Therefore we obtain

$$\operatorname*{ess\ sup}_{t\in[0,T]} \left(\widehat{u}(t)^T \widehat{u}(t)\right)^{1/2} = -x_0^T y = \int_0^T \widehat{u}(t)^T Y(t)^T \,\mathrm{d}t\, y$$

which is only possible, if

$$\widehat{u}(t) = \frac{-x_0^T y}{\left(y^T Y(t) Y(t)^T y\right)^{1/2}}\, Y(t)^T y \qquad \text{for almost all} \quad t \in [0,T]$$

This implies

$$\left(\widehat{u}(t)^T \widehat{u}(t)\right)^{1/2} = \varphi(\widehat{u}) = \varphi(u_T) = \left(u_T(t)^T u_T(t)\right)^{1/2} \quad \text{for almost all} \quad t \in [0,T].$$

Since $w(t) = \frac{1}{2}(\widehat{u}(t) + u_T(t))$, $t \in [0,T]$, is also a solution of the minimum norm problem, it follows

$$||\widehat{u}(t)||_2 = ||u_T(t)||_2 = ||w(t)||_2 \qquad \text{for almost all} \quad t \in [0,T]$$

which is only possible, if

$$\widehat{u}(t) = u_T(t) = w(t) \qquad \text{for almost all} \quad t \in [0,T].$$

This completes the proof. □

Now we make a first step towards a characterization of time-minimal null-controls by proving the

Theorem 2.12. *If restricted null-controllability holds and if the normality condition is satisfied, then the minimum time $T(\gamma)$ defined by (2.32) is also given by*

$$T(\gamma) = \inf\{T > 0 \mid v(T) \le \gamma\} \tag{2.44}$$

where

$$v(T) = \max\left\{ -x_0^T y \;\middle|\; \int_0^T \left(y^T Y(t) Y(t)^T y\right)^{1/2} \mathrm{d}t \le 1 \right\} \tag{2.45}$$

Proof. Let us put

$$\widehat{T}(\gamma) = \inf\{T > 0 \mid v(T) \le \gamma\}.$$

Then it is clear that $\widehat{T}(\gamma) \le T(\gamma)$ because of

$$\{T > 0 \mid (2.21) \text{ is satisfied for some } u \in \mathcal{U}\ (2.31)\} \subseteq \{T > 0 \mid v(T) \le \gamma\}$$

which is a consequence of (2.39).

If $\widehat{T}(\gamma) < T(\gamma)$, then by, Theorem 2.11, for every $T \in \left(\widehat{T}(\gamma), T(\gamma)\right)$ there is some $u_T \in \mathcal{U}$ which satisfies (2.21) and for which $\varphi(u_T) = v(T) \le \gamma$ holds true. This, however, contradicts the definition (2.32) of $T(\gamma)$ which completes the proof. □

Theorem 2.11 can be strengthened to

Theorem 2.13. *In addition to the assumption of Theorem 2.12 let the function $T \to v(t)$ (2.45) from $(0,\infty)$ into $(0,\infty)$ be continuous. Then it follows that*

$$v(T(\gamma)) = \gamma \tag{2.46}$$

and for each $u_\gamma \in \mathcal{U}$ with (2.33) it follows that $\varphi(u_\gamma) = v(T(\gamma))$, i.e., every time-minimal null-control is a minimum norm control of $[0, T(\gamma)]$ (we tacitly assume that $x_0 \ne \Theta_n$).

Proof. $x_0 \neq \Theta_n$ implies $T(\gamma) > 0$ and $\lim_{T\to 0+} v(T) = \infty$.

By virtue of the continuity of $T \to v(T)$ it follows from (2.44) that $v(T(\gamma)) \leq \gamma$. Let us assume that $v(T(\gamma)) < \gamma$. By the intermediate value theorem for continuous functions then it follows that there is some $T^* \in (T, T(\gamma))$ (where $v(T) > \gamma$) with $v(T^*) = \gamma$. This implies the existence of $u_{T^*} \in \mathcal{U}$ which satisfies (2.21) and $\varphi(u_{T^*}) = v(T^*) = \gamma$. This is a contradiction to the definition (2.32) of $T(\gamma)$. Hence (2.46) must be true from which the last assertions follow immediately. □

In view of Theorem 2.11 we have the following

Corollary 2.14. *Under the assumptions of Theorem 2.13 there is exactly one time-minimal control on* $[0, T(\gamma)]$ *which is the minimum norm control* $u_{T(\gamma)}(t) \in \mathcal{U}$ *on* $[0, T(\gamma)]$ *and is given by*

$$u_{T(\gamma)}(t) = \frac{\gamma}{\left(y_{T(\gamma)}^T Y(t)Y(t)^T y_{T(\gamma)}\right)^{1/2}} Y(t)^T y_{T(\gamma)} \qquad (2.47)$$

$$\text{for almost all} \quad t \in [0, T(\gamma)]$$

where $y_{T(\gamma)} \in \mathbb{R}^n$ *is a solution of the dual problem to the minimum norm problem for* $T = T(\gamma)$.

The continuity assumption for the function $T \to v(T)$ in Theorem 2.13 can be dispensed with, since it is a consequence of the normality condition.

In order to show this we at first observe that in the case $x_0 \neq \Theta_n$, for every $T > 0$, $v(T)$ defined by (2.45) is also given by $v(T) = 1/\lambda(T)$ where

$$\lambda(T) = \min\left\{ \int_0^T \left(y^T Y(t)Y(t)^T y\right)^{1/2} \mathrm{d}t \;\middle|\; -x_0^T y = 1 \right\} \qquad (2.48)$$

Theorem 2.15. *If the normality condition holds, then the function* $T \to \lambda(T)$ *(2.48),* $T > 0$, *is strictly increasing and continuous.*

Proof. Let $T_2 > T_1$ be given. Let

$$\lambda(T_k) = \int_0^{T_k} \left(y_{T_k}^T Y(t)Y(t)^T y_{T_k}\right)^{1/2} \mathrm{d}t \qquad \text{for} \quad k = 1, 2$$

with $-x_0^T y_{T_k} = 1$. For brevity we put, for every $y \in \mathbb{R}^n$,

$$||Y(t)^T y||_2 = \left(y^T Y(t)Y(t)^T y\right)^{1/2}, \qquad t \in \mathbb{R}.$$

At first we get

$$\begin{aligned}\lambda(T_1) &= \int_0^{T_1} ||Y(t)^T y_{T_1}||_2 \, dt \\ &\leq \int_0^{T_1} ||Y(t)^T y_{T_2}||_2 \, dt \leq \int_0^{T_2} ||Y(t)^T y_{T_2}||_2 \, dt = \lambda(T_2).\end{aligned}$$

It even follows that

$$\begin{aligned}\lambda(T_2) &= \int_0^{T_2} ||Y(t)^T y_{T_2}||_2 \, dt \\ &= \int_0^{T_1} ||Y(t)^T y_{T_2}||_2 \, dt + \int_{T_1}^{T_2} ||Y(t)^T y_{T_2}||_2 \, dt \\ &\geq \lambda(T_1) + \int_{T_1}^{T_2} ||Y(t)^T y_{T_2}||_2 \, dt\end{aligned}$$

where

$$\int_{T_1}^{T_2} ||Y(t)^T y_{T_2}||_2 \, dt > 0,$$

since otherwise $Y(t)\, y_{T_2} = \Theta_m$ for all $t \in [T_1, T_2]$ which, by the normality condition, implies $y_{T_2} = \Theta_n$ and contradicts $-x_0^T y_{T_2} = 1$.

As a result we obtain $\lambda(T_2) > \lambda(T_1)$ which shows that $T \to \lambda(T)$ is strictly increasing. On the other hand we also have

$$\begin{aligned}\lambda(T_2) &= \int_0^{T_2} ||Y(t)^T y_{T_2}||_2 \, dt \leq \int_0^{T_2} ||Y(t)^T y_{T_1}||_2 \, dt \\ &= \int_0^{T_1} ||Y(t)^T y_{T_1}||_2 \, dt + \int_{T_1}^{T_2} ||Y(t)^T y_{T_1}||_2 \, dt \\ &= \lambda(T_1) + \int_{T_1}^{T_2} ||Y(t)^T y_{T_1}||_2 \, dt,\end{aligned}$$

hence

$$0 \leq \lambda(T_2) - \lambda(T_1) \leq \int_{T_1}^{T_2} ||Y(t)^T y_{T_1}||_2 \, dt$$

which implies

$$\lim_{T_2 \to T_1 + 0} \lambda(T_2) = \lambda(T_1)$$

and shows the right continuity of the function $T \to \lambda(T)$.

The proof of the left continuity requires some preparation.

We choose $T_0 \in (0, T_1)$ arbitrarily and consider the function

$$y \to \int_0^{T_0} ||Y(t)^T y||_2 \, dt$$

from $\mathbb{R}^n$ into $(0, \infty)$ for which

$$m_{T_0} = \inf \left\{ \int_0^{T_0} ||Y(t)^T y||_2 \, dt \,\middle|\, ||y||_2 = \left(y^T y\right)^{1/2} \right\} > 0$$

holds true. This implies

$$m_{T_0} \, ||y||_2 \le \int_0^{T_0} ||Y(t)^T y||_2 \, dt \qquad \text{for all} \quad y \in \mathbb{R}^n.$$

In particular we obtain

$$\begin{aligned} m_{T_0} \, ||y_{T_1}||_2 &\le \int_0^{T_0} ||Y(t)^T y_{T_1}||_2 \, dt \\ &\le \int_0^{T_1} ||Y(t)^T y_{T_1}||_2 \, dt = \lambda(T_1) \le \lambda(T_2) \end{aligned}$$

and further

$$\begin{aligned} 0 \le \lambda(T_2) - \lambda(T_1) &\le \int_{T_1}^{T_2} ||Y(t)^T y_{T_1}||_2 \, dt \\ \le \int_{T_1}^{T_2} ||Y(t)^T||_2 \, dt \, ||y_{T_1}||_2 &\le \frac{\lambda(T_2)}{m_{T_0}} \int_{T_1}^{T_2} ||Y(t)^T||_2 \, dt \end{aligned}$$

with

$$||Y(t)^T||_2 = \left(\sum_{j,k=1}^{n} y_{jk}(t)^2 \right)^{1/2} .$$

From this we infer

$$\lim_{T_1 \to T_2 - 0} \lambda(T_1) = \lambda(T_2)$$

which shows the left-continuity of the function and completes the proof. □

Under the assumptions of Theorem 2.15 it follows that the function $T \to v(T) = 1/\lambda(T)$, $T > 0$, is strictly decreasing and continuous.

Thus we can strengthen Theorem 2.13 to

Theorem 2.16. *If restricted null-controllability holds and if the normality condition is satisfied, then the equation (2.46) for the minimum time $T(\gamma)$ holds true and there is exactly one time-minimal control on $[0, T(\gamma)]$ which is the unique minimum norm control $u_{T(\gamma)} \in L^\infty(\mathbb{R}, \mathbb{R}^m)$ on $[0, T(\gamma)]$ and is given by (2.47).*

2.2 The Time-Discrete Autonomous Case

2.2.1 The Problem of Fixed Point Controllability

We begin with a system of difference equations of the form

$$x(t+1) = g(x(t), u(t)), \qquad t \in \mathbb{N}_0 \tag{2.49}$$

where $g : \mathbb{R}^n \times \mathbb{R}^m \to \mathbb{R}^n$ is a continuous mapping.

The functions $x : \mathbb{N}_0 \to \mathbb{R}^n$ and $u : \mathbb{N}_0 \to \mathbb{R}^m$ are considered as state and control functions, respectively. For every control function $u : \mathbb{N}_0 \to \mathbb{R}^m$ and every vector $x_0 \in \mathbb{R}^n$ there exists exactly one state function $x : \mathbb{N}_0 \to \mathbb{R}^n$ which satisfies (2.49) and

$$x(0) = x_0. \tag{2.50}$$

If we fix the control function $u : \mathbb{N}_0 \to \mathbb{R}^m$ and define

$$f_{t+1}(x) = g(x, u(t)), \qquad x \in \mathbb{R}^n,\ t \in \mathbb{N}_0, \tag{2.51}$$

then, for every $t \in \mathbb{N}_0$, $f_t : \mathbb{R}^n \to \mathbb{R}^n$ is a continuous mapping and $\left(\mathbb{R}^n, (f_t)_{t\in\mathbb{N}}\right)$ is a non-autonomous time-discrete dynamical system which is controlled by the function $u : \mathbb{N}_0 \to \mathbb{R}^m$. If

$$u(t) = \Theta_m = \text{ zero vector of } \mathbb{R}^m \text{ for all } t \in \mathbb{N}_0,$$

then the system (2.49) is called uncontrolled. Let us assume that the uncontrolled system (2.49) admits fixed points $\widehat{x} \in \mathbb{R}^n$ which then solve the equation

$$\widehat{x} = g\left(\widehat{x}, \Theta_m\right). \tag{2.52}$$

Now let $\Omega \subseteq \mathbb{R}^m$ be a subset with $\Theta_m \in \Omega$. Then we define the set of admissible control functions by

$$U = \{u : \mathbb{N}_0 \to \mathbb{R}^m \mid u(t) \in \Omega \quad \text{for all} \quad t \in \mathbb{N}_0\}. \tag{2.53}$$

After these preparations we are in the position to formulate the

<u>Problem of Fixed Point Controllability:</u> Given a fixed point $\widehat{x} \in \mathbb{R}^n$ of the system

$$x(t+1) = g\left(x(t), \Theta_m\right), \qquad t \in \mathbb{N}_0, \tag{2.54}$$

i.e., a solution $\widehat{x}$ of equation (2.52) and an initial state $x_0 \in \mathbb{R}^n$ find some $N \in \mathbb{N}_0$ and a control function $u \in U$ with

$$u(t) = \Theta_m \qquad \text{for all} \quad t \geq N \tag{2.55}$$

such that the solution $x : \mathbb{N}_0 \to \mathbb{R}^n$ of (2.49), (2.50) satisfies the end condition

$$x(N) = \widehat{x}. \tag{2.56}$$

(This implies $x(t) = \widehat{x}$ for all $t \geq N$.) From (2.49) and (2.50) it follows that

$$x(N) = \underbrace{g(g(\cdots(g}_{N-\text{times}}(x_0, u(0)), u(1)), \cdots), u(N-1)) = G^N(x_0, u(0), \ldots, u(N-1)). \tag{2.57}$$

Let $N \in \mathbb{N}$ be given. If $u(0), \ldots, u(N-1) \in \Omega$ are solutions of the system

$$G^N(x_0, u(0), \ldots, u(N-1)) = \widehat{x} \tag{2.58}$$

and one defines

$$u(t) = \Theta_m \qquad \text{for all} \quad t \geq N,$$

then one obtains a control function $u : \mathbb{N}_0 \to \mathbb{R}^m$ which solves the problem of fixed point controllability.

From the definition (2.57) it follows that

$$G^N(x_0, u(0), \ldots, u(N-1)) = G^1\left(G^{N-1}(x_0, u(0), \ldots, u(N-2)), u(N-1)\right).$$

Conversely now let us assume that we are given a sequence $\left(G^N\right)_{N \in \mathbb{N}}$ of vector functions $G^N : \mathbb{R}^n \times \mathbb{R}^{m \cdot N} \to \mathbb{R}^n$ with this property.

Then we define, for every $t \in \mathbb{N}$,

$$x(t) = G^t(x_0, u(0), \ldots, u(t-1))$$
$$\text{for } x_0 \in \mathbb{R}^n \text{ and } u(s) \in \mathbb{R}^m \text{ for } s = 0, \ldots, t-1$$

and conclude

$$x(t) = G^1\left(G^{t-1}(x_0, u(0), \ldots, u(t-2)), u(t-1)\right) = G^1(x(t-1), u(t-1)) \quad \text{for all} \quad t \in \mathbb{N},$$

if we define $G^0(x_0) = x_0$. Now let $S(\widehat{x}) \subseteq \mathbb{R}^n$ be the set of all vectors $x_0 \in \mathbb{R}^n$ such that there exists a time $N \in \mathbb{N}$ and a solution $\left(u(0)^T, \ldots, u(N-1)^T\right)^T \in \Omega^N$ of the system (2.58). Obviously, $\widehat{x} \in S(\widehat{x})$ (choose $N = 1$ and $u(0) = \Theta_n$). A first simple sufficient condition for the solvability of the problem of fixed point controllability is then given by

Proposition 2.17. *Let $\widehat{x} \in S(\widehat{x})$ be an interior point of $S(\widehat{x})$ and let $\widehat{x}$ be an attractor of the uncontrolled system (2.54), i.e., there exists an open set $U \subseteq \mathbb{R}^n$ with $\widehat{x} \in U$ such that $\lim_{t \to \infty} x(t) = \widehat{x}$ where $x : \mathbb{N}_0 \to \mathbb{R}^n$ is a solution of (2.54) with (2.50) for any $x_0 \in U$. Then it follows that $S(\widehat{x}) \supseteq U$ which implies that for every choice of $x_0 \in U$ the problem of fixed point controllability has a solution.*

Proof. Since $\widehat{x} \in S(\widehat{x})$ is an interior point of $S(\widehat{x})$, there is an open neighborhood $W(\widehat{x}) \subseteq \mathbb{R}^n$ of $\widehat{x}$ with $W(\widehat{x}) \subseteq S(\widehat{x})$.

Now let $x_0 \in U$ be chosen arbitrarily. Then there is some $N_1 \in \mathbb{N}$ with $x(N_1) \in W(\widehat{x})$ where $x : \mathbb{N}_0 \to \mathbb{R}^m$ is a solution of (2.54), (2.50). This implies the existence of some $N_2 \in \mathbb{N}$ and a solution $\left(u(0)^T, \ldots, u(N_2-1)^T\right)^T \in \Omega^{N_2}$ with

$$G^{N_2}(x(N_1), u(0), \ldots, u(N_2-1)) = \widehat{x}.$$

If we define

$$u^*(t) = \begin{cases} \Theta_m & \text{for} \quad t = 0, \ldots, N_1 - 1, \\ u(t - N_1) & \text{for} \quad t = N_1, \ldots, N_1 + N_2 - 1, \end{cases}$$

then it follows that

$$G^N(x_0, u^*(0), \ldots, u^*(N-1)) = \widehat{x}$$

where $N = N_1 + N_2$, i.e., $x_0 \in S(\widehat{x})$ which completes the proof. □

The essential assumption in Proposition 2.17 is that the fixed point $\widehat{x}$ of the uncontrolled system (2.54) is an interior point of the controllable set $S(\widehat{x})$.

In order to find sufficient conditions for that we assume that Ω is open and $g \in C^1(\mathbb{R}^n \times \mathbb{R}^m)$ which implies $G^N \in C^1\left(\mathbb{R}^n \times \mathbb{R}^{m \cdot N}\right)$ for every $N \in \mathbb{N}$ and

$$\begin{aligned} & G_x^N(x, u(0), \ldots, u(N-1)) \\ & \quad = g_x\left(G^{N-1}(x_0, u(0), \ldots, u(N-2)), u(N-1)\right) \\ & \qquad \times g_x\left(G^{N-2}(x_0, u(0), \ldots, u(N-3)), u(N-2)\right) \\ & \qquad \times \qquad \vdots \\ & \qquad \times g_x(x, u(0)), \end{aligned}$$

and

$$\begin{aligned} & G_{u(k)}^N(x, u(0), \ldots, u(N-1)) \\ & \quad = g_x\left(G^{N-1}(x_0, u(0), \ldots, u(N-2)), u(N-1)\right) \\ & \qquad \times g_x\left(G^{N-2}(x_0, u(0), \ldots, u(N-3)), u(N-2)\right) \\ & \qquad \times \qquad \vdots \\ & \qquad \times g_x\left(G^{k+1}(x, u(0), \ldots, u(k)), u(k+1)\right) \\ & \qquad \times g_{u(k)}\left(G^k(x, u(0), \ldots, u(k-1)), u(k)\right) \end{aligned}$$

for $k = 0, \ldots, N-1$.

Let us assume that $g_x(\widehat{x}, \Theta_m)$ is non-singular. Then it follows, for all $N \in \mathbb{N}$, that $G_x^N(\widehat{x}, \Theta_m, \ldots, \Theta_m)$ is also non-singular, since

$$G_x^N(\widehat{x}, \Theta_m, \ldots, \Theta_m) = g_x(\widehat{x}, \Theta_m)^N .$$

Since

$$\widehat{x} = G^N(\widehat{x}, \Theta_m, \ldots, \Theta_m),$$

there exists, by the implicit function theorem, an open set $V \subseteq \Omega^N$ with $\Theta_m^N \in V$ and a function $h : V \to \mathbb{R}^n$ with $h \in \mathcal{C}^1(V)$ such that

$$h\left(\Theta_m^N\right) = \widehat{x} \quad \text{and} \quad G^N(h(u(0), \ldots, u(N-1)), u(0), \ldots, u(N-1)) = \widehat{x}$$

for all $(u(0), \ldots, u(N-1)) \in V$.

Moreover,

$$\begin{aligned} h_{u(k)}\left(\Theta_m^N\right) &= -G_x^N(\widehat{x}, \Theta_m, \ldots, \Theta_m)^{-1} G_{u(k)}(\widehat{x}, \Theta_m, \ldots, \Theta_m) \\ &= -g_x(\widehat{x}, \Theta_m)^{-N} g_x(\widehat{x}, \Theta_m)^{N-k} g_{u(k)}(\widehat{x}, \Theta_m) \\ &= -g_x(\widehat{x}, \Theta_m)^{-k} g_{u(k)}(\widehat{x}, \Theta_m) \end{aligned}$$

for $k = 0, \ldots, N-1$.

Result. *If $g_x(\widehat{x}, \Theta_m)$ is non-singular, then, for every $N \in \mathbb{N}$, there is an open set $V_N \subseteq \Omega^N$ with $\Theta_m^N \in V_N$ and a function $h_N : V_N \to \mathbb{R}^n$ with $h_N \in \mathcal{C}^1(V_N)$ such that*

$$h_N(u(0), \ldots, u(N-1)) \in S(\widehat{x}) \qquad \textit{for all} \quad (u(0), \ldots, u(N-1)) \in V_N.$$

We next assume that, for some $N \in \mathbb{N}$,

$$\operatorname{rank}\left(h_{u(0)}\left(\Theta_m^N\right) \mid \ldots \mid h_{u(N-1)}\left(\Theta_m^N\right)\right) = n.$$

Then there are n columns in the $n \times m \cdot N$–matrix

$$\left(h_{u(0)}\left(\Theta_m^N\right) \mid \ldots \mid h_{u(N-1)}\left(\Theta_m^N\right)\right)$$

which are linearly independent.

Now let E be the n–dimensional subspace of $\mathbb{R}^{m \cdot N}$ consisting of all vectors whose components vanish that do not correspond to the above linearly independent columns.

If we put $U = E \cap V_N$, then U is an open subset of E and the restriction of h to U is a $\mathcal{C}^1$–function on U whose Jacobi matrix at Θ_m^N consists of the above linearly independent columns of $\left(h_{u(0)}\left(\Theta_m^N\right) \mid \ldots \mid h_{u(N-1)}\left(\Theta_m^N\right)\right)$ and is therefore invertible. By the inverse function theorem there exist open sets $\widetilde{U} \subseteq E \cap V_N$ and $\widetilde{V} \subseteq \mathbb{R}^n$ with $\Theta_m^N \in \widetilde{U}$ and $\widehat{x} \in \widetilde{V}$ such that h is homeomorphic on $\widetilde{U}$ and $h\left(\widetilde{U}\right) = \widetilde{V}$.

This implies that $\widehat{x}$ is an interior point of $S(\widehat{x})$.

Let us demonstrate this result by the predator prey model that was investigated with respect to asymptotical stability in Section 1.3.7. We consider this model as a controlled system of the form

$$\begin{aligned} x_1(t+1) &= x_1(t) + a\,x_1(t) - e\,x_1(t)^2 - b\,x_1(t)x_2(t) - x_1(t)\,u_1(t), \\ x_2(t+1) &= x_2(t) - c\,x_2(t) + d\,x_1(t)x_2(t) - x_2(t)\,u_2(t), \qquad t \in \mathbb{N}_0. \end{aligned}$$

The mapping $g : \mathbb{R}^2 \times \mathbb{R}^2 \to \mathbb{R}^2$ in (2.49) is therefore given by

$$\begin{pmatrix} (1+a)\,x_1 - e\,x_1^2 - b\,x_1x_2 - x_1\,u_1 \\ (1-c)\,x_2 + d\,x_1x_2 - x_2\,u_2 \end{pmatrix}, \qquad x, u \in \mathbb{R}^2.$$

We have seen in Section 1.3.7 that

$$\widehat{x} = \left(\frac{c}{d}, \frac{1}{b}\left(a - \frac{ce}{d}\right)\right)^T$$

is the only fixed point of the uncontrolled system (with $u_1 = u_2 = 0$) in $\overset{\circ}{\mathbb{R}}{}^2_+ \times \overset{\circ}{\mathbb{R}}{}^2_+$, if $a > ce/d$.

One calculates

$$g_x(\widehat{x}, \Theta_2) = \begin{pmatrix} 1 - e\dfrac{c}{d} & -\dfrac{bc}{d} \\ \dfrac{d}{b}\left(a - \dfrac{ce}{d}\right) & 1 \end{pmatrix}$$

which implies that $g_x(\widehat{x}, \Theta_2)$ is non-singular, if and only if

$$1 - e\frac{c}{d} + c\left(a - \frac{ce}{d}\right) \neq 0. \qquad (*)$$

Further we obtain

$$g_u(\widehat{x}, \Theta_2) = \begin{pmatrix} -\dfrac{c}{d} & 0 \\ 0 & -\dfrac{1}{b}\left(a - \dfrac{ce}{d}\right) \end{pmatrix}$$

which implies

$$\operatorname{rank}\left(h_{u(0)}(\Theta_2)\right) = \operatorname{rank}\left(-g_{u(0)}(\widehat{x}, \Theta_2)\right) = 2.$$

Hence $\widehat{x}$ is an interior point of $S(\widehat{x})$, if $(*)$ is satisfied.

This example is a special case of the following situation:

Let

$$g(x,u) = f(x) + F(x)\,u = f(x) + \sum_{i=1}^{m} f_i(x)\,u_i,$$
$$x \in \mathbb{R}^n,\ u_1, \ldots, u_m \in \mathbb{R}, \text{ where } f, f_i \in C^1(\mathbb{R}^n),\ i = 1, \ldots, m.$$

Let $\widehat{x} \in \mathbb{R}^n$ be a fixed point of f, i.e.,

$$\widehat{x} = f(\widehat{x}) = g(\widehat{x}, \Theta_m).$$

Then,

$$g_x(\widehat{x}, \Theta_m) = f_x(\widehat{x}) \qquad \text{and} \qquad g_u(\widehat{x}, \Theta_m) = F(\widehat{x})$$

and hence

$$\begin{aligned}&\left(h_{u(0)}\left(\Theta_m^N\right) \mid \ldots \mid h_{u(N-1)}\left(\Theta_m^N\right)\right)\\&\quad = -\Big(F(\widehat{x}) \mid f_x(\widehat{x})^{-1} F(\widehat{x}) \mid \ldots \mid f_x(\widehat{x})^{-N+1} F(\widehat{x})\Big).\end{aligned}$$

In the example we have $m = n = 2$ and the 2×2–matrix $F(\widehat{x})$ is non-singular. Next we come back to the solution of (2.58) which we replace by an optimization problem. For this purpose we define a cost functional $\varphi : \mathbb{R}^{m \cdot N} \to \mathbb{R}$ by putting

$$\begin{aligned}&\varphi(u(0), \ldots, u(N-1)) = \left\| G^N(x_0, u(0), \ldots, u(N-1)) - \widehat{x}\right\|_2^2\\&\quad \text{for} \quad (u(0), \ldots, u(N-1)) \in \mathbb{R}^{m \cdot N} \qquad (\|\cdot\|_2 = \text{Euclidian norm in } \mathbb{R}^n)\end{aligned}$$

and try to find $(u(0), \ldots, u(N-1)) \in \Omega^N$ such that $\varphi(u(0), \ldots, u(N-1)) \leq \varphi(\widetilde{u}(0), \ldots, \widetilde{u}(N-1))$ for all $(\widetilde{u}(0), \ldots, \widetilde{u}(N-1)) \in \Omega^N$.

If $\varphi(u(0), \ldots, u(N-1)) = 0$, then $\varphi(u(0), \ldots, u(N-1)) \in \Omega^N$ solves the equation (2.58). Otherwise no such solution exists. We again assume that $g \in C^1(\mathbb{R}^n \times \mathbb{R}^m)$. Let $\Omega \subseteq \mathbb{R}^n$ be open.

Then a necessary condition for $u(0), \ldots, u(N-1)) \in \Omega^N$ to minimize φ on Ω^N is given by

$$\begin{aligned}\varphi_{u(k)}(u(0), \ldots, u(N-1)) &= 2\, G_{u(k)}^N(x_0, u(0), \ldots, u(N-1))^T\\&\quad \times \left(G^N(x_0, u(0), \ldots, u(N-1)) - \widehat{x}\right)\\&= \Theta_m \qquad \text{(OC)}\end{aligned}$$

for all $k = 0, \ldots, N-1$.

For the determination of $(u(0), \ldots, u(N-1)) \in \Omega^N$ with (OC) one can apply Marquardt's algorithm: Let $(u(0), \ldots, u(N-1)) \in \Omega^N$ be chosen. If (OC) is satisfied, then $(u(0), \ldots, u(N-1))$ is taken as a solution of the optimization problem. Otherwise, for every $k \in \{0, \ldots, N-1\}$, a vector $h_\lambda(k) \in \mathbb{R}^m$ is determined as a solution of the linear system

$$\begin{aligned}&\Big(2\, G_{u(k)}^N(x_0, u(0), \ldots, u(N-1))\Big)^T G_{u(k)}^N(x_0, u(0), \ldots, u(N-1)) + \lambda\, I_m\, h_\lambda(k)\\&\quad = 2\, G_{u(k)}^N(x_0, u(0), \ldots, u(N-1))^T \Big(G_{u(k)}^N(x_0, u(0), \ldots, u(N-1)) - \widehat{x}\Big)\end{aligned}$$

where $\lambda > 0$ and I_m is the $m \times m$–unit matrix.

Then one can show (see, for instance [22]) that for sufficiently large $\lambda > 0$ it follows that

$$(u(0) + h_\lambda(0), \ldots, u(N-1) + h_\lambda(N-1)) \in \Omega$$

and

$$\varphi(u(0) + h_\lambda(0), \ldots, u(N-1) + h_\lambda(N-1)) < \varphi(u(0), \ldots, u(N-1)).$$

The algorithm is then continued with $(u(0) + h_\lambda(0), \ldots, u(N-1) + h_\lambda(N-1))$ instead of $(u(0), \ldots, u(N-1))$.

Special Case: Modelling of Conflicts

Now let us consider a special case which is motivated by a situation which occurs in the modelling of conflicts. We begin with an uncontrolled system of the form

$$\begin{aligned} x^1(t+1) &= g_1\left(x^1(t), x^2(t)\right), \\ x^2(t+1) &= g_2\left(x^1(t), x^2(t)\right), \qquad t \in \mathbb{N}_0, \end{aligned}$$

where $g_i : \mathbb{R}^{n_1} \times \mathbb{R}^{n_2} \to \mathbb{R}^{n_i}$, $i = 1, 2$, are given continuous mappings and $x^i : \mathbb{N}_0 \to \mathbb{R}^{n_i}$, $i = 1, 2$, are considered as state functions.

For $t = 0$ we assume initial conditions

$$x^1(0) = x_0^1, \qquad x^2(0) = x_0^2 \tag{2.59}$$

where $x_0^1 \in \mathbb{R}^{n_1}$ and $x_0^2 \in \mathbb{R}^{n_2}$ are given vectors with

$$\Theta_{n_2} \leq x_0^2 \leq x_*^2$$

for some $x_*^2 \geq \Theta_{n_2}$ which is also given. We further assume that the above system admits fixed points $\left(\widehat{x}^{1T}, \widehat{x}^{2T}\right)^T \in \mathbb{R}^{n_1} \times \mathbb{R}^{n_2}$ with

$$\Theta_{n_2} \leq \widehat{x}^2 \leq x_*^2$$

which are then solutions of the system

$$\widehat{x}^1 = g_1\left(\widehat{x}^1, \widehat{x}^2\right), \qquad \widehat{x}^2 = g_2\left(\widehat{x}^1, \widehat{x}^2\right).$$

Now we consider the following

Problem: Find vector functions $x^1 : N_0 \to \mathbb{R}^{n_1}$ and $x^2 : \mathbb{N}_0 \to \mathbb{R}^{n_2}$ with

$$\Theta_{n_2} \le x^2(t) \le x^2_* \qquad \text{for all} \quad t \in \mathbb{N}_0$$

which satisfy the above system equations and initial conditions and

$$x^1(t) = \widehat{x}^1, \quad x^2(t) = \widehat{x}^2 \qquad \text{for all} \quad t \ge N$$

where $N \in \mathbb{N}_0$ is a suitably chosen integer. In general this problem will not have a solution. Therefore we replace the uncontrolled system by the following controlled system

$$\begin{aligned} x^1(t+1) &= g_1\left(x^1(t), x^2(t) + u(t)\right), \\ x^2(t+1) &= g_2\left(x^1(t), x^2(t) + u(t)\right), \qquad t \in \mathbb{N}_0, \end{aligned} \tag{2.60}$$

where $u : \mathbb{N}_0 \to \mathbb{R}^{n_2}$ is a control function. Then we consider the problem of finding a control function $u : \mathbb{N}_0 \to \mathbb{R}^{n_2}$ such that the solutions $x^1 : \mathbb{N}_0 \to \mathbb{R}^{n_1}$ and $x^2 : \mathbb{N}_0 \to \mathbb{R}^{n_2}$ of (2.60) and (2.59) satisfy the conditions

$$\Theta_{n_2} \le x^2(t) + u(t) \le x^2_* \qquad \text{for all} \quad t \in \mathbb{N}_0$$

and

$$x^1(t) = \widehat{x}^1, \quad x^2(t) + u(t) = \widehat{x}^2 \qquad \text{for all} \quad t \ge N$$

where $N \in \mathbb{N}_0$ is a suitably chosen integer.

Let us assume that we can find a vector function $v : \mathbb{N}_0 \to \mathbb{R}^n$ with

$$\Theta_{n_2} \le v(t) \le x^2_* \quad \text{for} \quad t = 0, \ldots, N-1 \qquad \text{and} \qquad v(t) = \widehat{x}^2 \quad \text{for} \quad t \ge N$$

such that the solution $x^1 : \mathbb{N}_0 \to \mathbb{R}^{n_1}$ of

$$\begin{aligned} x^1(0) &= x^1_0, \\ x^1(t+1) &= g_1\left(x^1(t), v(t)\right), \qquad t \in \mathbb{N}_0 \end{aligned}$$

satisfies

$$x^1(t) = \widehat{x}^1 \qquad \text{for all} \quad t \ge N$$

where $N \in \mathbb{N}$ is a suitably chosen integer.

Then we put

$$\begin{aligned} x^2(0) &= x^2_0, \\ x^2(t+1) &= g_2\left(x^1(t), v(t)\right), \qquad t \in \mathbb{N}_0 \end{aligned}$$

and define

$$u(t) = v(t) - x^2(t) \qquad \text{for} \quad t \ge N_0.$$

With these definitions we obtain a solution of the above control problem.

Thus, in order to find such a solution, we have to find a vector function $v : \mathbb{N}_0 \to \mathbb{R}^{n_2}$ with

$$\begin{aligned} \Theta_{n_2} \leq v(t) \leq x_*^2 \quad & \text{for} \quad t = 0, \ldots, N-1, \\ v(t) = \widehat{x}^2 \quad & \text{for} \quad t \geq N \end{aligned}$$

such that the solution $x^1 : \mathbb{N}_0 \to \mathbb{R}^{n_1}$ of

$$\begin{aligned} x^1(0) &= x_0^1, \\ x^1(t+1) &= g_1\left(x^1(t), v(t)\right), \qquad t \in \mathbb{N}_0 \end{aligned}$$

satisfies

$$x^1(t) = \widehat{x}^1 \qquad \text{for all} \quad t \geq N$$

where $N \in \mathbb{N}$ is a suitably chosen integer. Let us demonstrate all this by an emission reduction model (1.65) to which we add the conditions

$$0 \leq M_i(t) \leq M_i^* \qquad \text{for all} \quad t \in \mathbb{N}_0 \quad \text{and} \quad i = 1, \ldots, r$$

and the initial conditions

$$E_i(0) = E_{0i} \quad \text{and} \quad M_i(0) = M_{0i} \qquad \text{for} \quad i = 1, \ldots, r$$

where $E_{0i} \in \mathbb{R}$ and $M_{0i} \in \mathbb{R}$ with $0 \leq M_{0i} \leq M_i^*$ for $i = 1, \ldots, r$ are given.

The corresponding controlled system (2.60) reads in this case

$$\begin{aligned} E_i(t+1) &= E_i(t) + \sum_{j=1}^{r} em_{ij}\left(M_j(t) + u_j(t)\right), \\ M_i(t+1) &= M_i(t) + u_i(t) - \lambda_i\left(M_i(t) + u_i(t)\right)\left(M_i^* - M_i(t) - u_i(t)\right) E_i(t) \end{aligned}$$

for $i = 1, \ldots, r$ and $t \in \mathbb{N}_0$.

The control functions $u_i : \mathbb{N}_0 \to \mathbb{R}$, $i = 1, \ldots, r$, must satisfy the conditions

$$0 \leq M_i(t) + u_i(t) \leq M_i^* \qquad \text{for} \quad i = 1, \ldots, r \quad \text{and} \quad t \in \mathbb{N}_0.$$

Fixed points of the system (1.65) are of the form $\left(\widehat{E}^T, \widehat{\Theta}_r^T\right)^T$ with $\widehat{E} \in \mathbb{R}^r$ arbitrary.

We have to find a vector function $v : \mathbb{N}_0 \to \mathbb{R}^r$ with

$$\begin{aligned} \Theta_r \le v(t) \le M^* &\quad \text{for } \ t = 0, \dots, N-1, \\ v(t) = \Theta_r &\quad \text{for } \ t \ge N \end{aligned}$$

such that the solution $E : \mathbb{N}_0 \to \mathbb{R}^r$ of

$$\begin{aligned} E(0) &= E_0, \\ E(t+1) &= E(t) + C\, v(t), \qquad t \in \mathbb{N}_0, \ \left(C = (em_{ij})_{i,j=1,\dots,r}\right) \end{aligned}$$

satisfies

$$E(t) = \widehat{E} \qquad \text{for all} \quad t \ge N$$

where $N \in \mathbb{N}$ is a suitably chosen integer.

First of all we observe that for every $N \in \mathbb{N}$

$$E(N) = E_0 + C\left(\sum_{t=0}^{N-1} v(t)\right).$$

Let us assume that C is invertible and C^{-1} is positive. Further we assume that $\widehat{E} \ge E_0$.

Then $E(N) = \widehat{E}$, if and only if

$$\sum_{t=0}^{N-1} v(t) = C^{-1}\left(\widehat{E} - E_0\right) \ge \Theta_r.$$

If we define

$$v(t) = \Theta_r \qquad \text{for all} \quad t \ge N,$$

then

$$E(t) = \widehat{E} \qquad \text{for all} \quad t \ge N.$$

Let us put

$$v_N = \sum_{t=0}^{N-1} v(t) = C^{-1}\left(\widehat{E} - E_0\right).$$

If we define

$$v(t) = \frac{1}{N} v_N \qquad \text{for} \quad t = 0, \dots, N-1$$

then

$$\sum_{t=0}^{N-1} v(t) = C^{-1}\left(\widehat{E} - E_0\right) \quad \text{and} \quad \Theta_r \le v(t) \le M^* \quad \text{for} \quad t = 0, \dots, N-1$$

for sufficiently large N, if $M_i^* > 0$ for all $i = 1, \dots, r$.

We finish with a numerical example: $r = 3$, $E_0 = (0,0,0)^T$, $\widehat{E} = (10,10,10)^T$, $M^* = (1,1,1)^T$, and

$$C = \begin{pmatrix} 1 & -0.8 & 0 \\ 0 & 1 & -0.8 \\ -0.1 & -0.5 & 1 \end{pmatrix}.$$

Then we have to solve the linear system

$$\begin{aligned} v_{N_1} - 0.8\, v_{N_2} \qquad\qquad &= 10, \\ v_{N_2} - 0.8\, v_{N_3} &= 10, \\ -0.1\, v_{N_1} - 0.5\, v_{N_2} + \qquad v_{N_3} &= 10. \end{aligned}$$

The solution reads

$$\begin{aligned} v_{N_1} &= 38.059701, \\ v_{N_2} &= 35.074627, \\ v_{N_3} &= 31.343284. \end{aligned}$$

We choose $N = 39$. Then we have to put

$$v(t) = \frac{1}{N}\, v_N = \begin{pmatrix} 0.9758898 \\ 0.8993494 \\ 0.803674 \end{pmatrix} \qquad \text{for} \quad t = 0, \ldots, 38.$$

2.2.2 Null-Controllability of Linear Systems

Instead of (2.49) we consider a linear system of the form

$$x(t+1) = A\,x(t) + B\,u(t), \qquad t \in \mathbb{N}_0, \tag{2.61}$$

where A is a real $n \times n$–matrix and B a real $n \times m$–matrix and where $u : \mathbb{N}_0 \to \mathbb{R}^m$ is a given control function. The corresponding uncontrolled system reads

$$x(t+1) = A\,x(t), \qquad t \in \mathbb{N}_0, \tag{2.62}$$

and admits $\widehat{x} = \Theta_n$ as a fixed point.

The problem of fixed point controllability is then equivalent to the

Problem of Null-Controllability: Given $x_0 \in \mathbb{R}^n$ find some $N \in \mathbb{N}_0$ and a control function $u \in U$ (2.53) with (2.55) such that the solution $x : \mathbb{N}_0 \to \mathbb{R}^n$ of (2.61), (2.50) satisfies the end condition

$$x(N) = \Theta_n \tag{2.63}$$

(which implies $x(t) = \Theta_n$ for all $t \geq N$).

From (2.61) and (2.50) it follows that

$$x(N) = A^N x_0 + \sum_{t=1}^{N} A^{N-t} B\, u(t-1) \tag{2.64}$$

so that (2.63) turns out to be equivalent to

$$\sum_{t=1}^{N} A^{N-t} B\, u(t-1) = -A^N x_0. \tag{2.65}$$

Now let A be non-singular. Then the set $S(\Theta_n)$ of all vectors $x_0 \in \mathbb{R}^n$ such that there exists a time $N \in \mathbb{N}$ and a solution $\left(u(0)^T, \ldots, u(N-1)^T\right)^T \in \Omega^N$ of the system (2.61) is given by

$$S(\Theta_n) = \bigcup_{N \in \mathbb{N}} E(N)$$

where, for every $N \in \mathbb{N}$,

$$E(N) = \left\{ x = \sum_{t=1}^{N} A^{N-t} B\, u(t-1) \;\middle|\; u \in U \text{ (2.53)} \right\}.$$

Next we assume that $\Omega \subseteq \mathbb{R}^m$ is convex, has Θ_m as interior point and satisfies

$$u \in \Omega \quad \Longrightarrow \quad -u \in \Omega. \tag{2.66}$$

Then, for every $N \in \mathbb{N}$, the set $E(N)$ is convex and $E(N) = -E(N)$. This implies because of

$$E(N) \subseteq E(N+1) \qquad \text{for all} \quad N \in \mathbb{N}$$

that $S(\Theta_n)$ is also convex and $S(\Theta_n) = -S(\Theta_n)$.

Further we assume Kalman's condition, i.e., there exists some $N_0 \in \mathbb{N}$ such that

$$\operatorname{rank}\left(B \mid AB \mid \ldots \mid A^{N_0-1}B\right) = n. \tag{2.67}$$

Then we can prove

Theorem 2.18. *If A is non-singular, $\Omega \subseteq \mathbb{R}^m$ is convex, has Θ_m as interior point and satisfies (2.66) and if Kalman condition (2.67) is satisfied for some $N_0 \in \mathbb{N}$, then Θ_n is an interior point of $S(\Theta_n)$.*

Proof. Let us assume that Θ_n is not an interior point of $S(\Theta_n)$. Then $S(\Theta_n)$ must be contained in a hyperplane through Θ_n, i.e., there must exist some $y \in \mathbb{R}^n$, $y \neq \Theta_n$, with

$$y^T x = 0 \qquad \text{for all} \quad x \in S(\Theta_n).$$

This implies

$$y^T \left(\sum_{t=1}^{N} A^{N-t} B\, u(t-1) \right) = 0$$

$$\text{for all} \quad \left(u(0)^T, \ldots, u(N-1)^T\right)^T \in \mathbb{R}^{m \cdot N} \quad \text{and all} \quad N \in \mathbb{N},$$

hence

$$y^T A^{N-t} B = \Theta_m^T \qquad \text{for all} \quad N \in \mathbb{N}.$$

In particular for $N = N_0$ this implies $y = \Theta_n$ due to the Kalman condition (2.67) which contradicts $y \neq \Theta_n$. Therefore the assumption that Θ_n is not an interior point of $S(\Theta_n)$ is false. □

In addition to the assumption of Theorem 2.18 we assume that all the eigenvalues of A are less than 1 in absolute value. Then according to the Corollary following Theorem 1.6 Θ_n is a global attractor of the uncontrolled system (2.62), i.e., $\lim_{t \to \infty} x(t) = \Theta_n$ where $x : \mathbb{N}_0 \to \mathbb{R}^n$ is a solution (2.62) with (2.50) for any $x_0 \in \mathbb{R}^n$. By Proposition 2.17 therefore the problem of null-controllability has a solution for every choice of $x_0 \in \mathbb{R}^n$. If the set Ω of control vector values has the form

$$\Omega = \{u \in \mathbb{R}^m \mid ||u|| \leq \gamma\} \tag{2.68}$$

for some $\gamma > 0$ where $|| \cdot ||$ is any norm in $\mathbb{R}^m$, then this result can be strengthened to

Theorem 2.19. *Let the Kalman condition (2.67) be satisfied for some $N_0 \in \mathbb{N}$. Further let Ω be of the form (2.68). Finally let all the eigenvalues of A^T be less than or equal to one in absolute value and the corresponding eigenvectors be linearly independent.*

Then the problem of null-controllability has a solution for every choice of $x_0 \in \mathbb{R}^n$, if A is non-singular.

Proof. We have to show that for every choice of $x_0 \in \mathbb{R}^n$ there is some $N \in \mathbb{N}_0$ and a control function $u \in U$ (2.53) such that (2.65) is satisfied. Since A is non-singular, (2.65) is equivalent to

$$\sum_{t=1}^{N} A^{-t} B\, u(t-1) = -x_0.$$

For every $N \in \mathbb{N}$ we define the convex set

$$R(N) = \left\{ x = \sum_{t=1}^{N} A^{-t} B\, u(t-1) \,\middle|\, u \in U \right\}$$

and put

$$R_\infty = \bigcup_{N \in \mathbb{N}} R(N).$$

Because of

$$R(N) \subseteq R(N+1) \qquad \text{for all} \quad N \in \mathbb{N}_0$$

the set R_∞ is also convex. We have to show that $R_\infty = \mathbb{R}^n$. Let us assume that $R_\infty \neq \mathbb{R}^n$.

Then there exists some $\widetilde{x} \in \mathbb{R}^n$ with $\widetilde{x} \notin R_\infty$ which can be separated from R_∞ by a hyperplane, i.e., there exists a number $\alpha \in \mathbb{R}$ and a vector $y \in \mathbb{R}^n$, $y \neq \Theta_n$ such that

$$y^T x \leq \alpha \leq y^T \widetilde{x} \qquad \text{for all} \quad x \in R_\infty.$$

Since $\Theta_n \in R_\infty$, it follows that $\alpha \geq 0$. Further it follows from the implication $u \in \Omega \;\Rightarrow\; -u \in \Omega$ that

$$\left| \sum_{t=1}^{N} y^T A^{-t} B\, u(t-1) \right| \leq \alpha \qquad \text{for all} \quad N \in \mathbb{N} \quad \text{and all} \quad u \in U.$$

This implies

$$\sum_{t=1}^{N} \left\| \left(y^T A^{-t} B \right)^T \right\|_d \leq \alpha \qquad \text{for all} \quad N \in \mathbb{N}$$

where $\|\cdot\|_d$ is the norm in $\mathbb{R}^m$ which is dual to $\|\cdot\|$. This in turn implies

$$\lim_{t \to \infty} y^T A^{-t} B = \Theta_m^T. \tag{2.69}$$

From the Kalman condition (2.67) it follows that there exist n linearly independent vectors in $\mathbb{R}^n$ of the form

$$c_i = A^{t_i}\, b_{j_i} \qquad \text{for} \quad i = 1, \ldots, n$$

where $t_i \in \{0, \ldots, N_0 - 1\}$ and $j_i \in \{1, \ldots, m\}$ and b_{j_i} denotes the j_i–th column vector of B. From (2.69) it follows that

$$\lim_{t \to \infty} y^T A^{-t} c_i = 0 \qquad \text{for} \quad i = 1, \ldots, n.$$

This implies

$$\lim_{t \to \infty} y^T A^{-t} = \Theta_n^T$$

or, equivalently,

$$\lim_{t\to\infty} \left(A^T\right)^{-t} y = \Theta_n. \tag{2.70}$$

Now let $\lambda_1, \ldots, \lambda_n \in \mathbb{C}$ be eigenvalues of A^T and $y_1, \ldots, y_n \in \mathbb{C}^n$ corresponding linearly independent eigenvectors. Then there is a unique representation

$$y = \sum_{j=1}^{n} \alpha_j y_j \qquad \text{where not all} \quad \alpha_j \in \mathbb{C} \quad \text{are zero}$$

and

$$\left(A^T\right)^{-t} y_j = \left(\frac{1}{\lambda_j}\right)^t y_j \qquad \text{for} \quad j = 1, \ldots, n,$$

hence

$$\left(A^T\right)^{-t} y = \sum_{j=1}^{n} \alpha_j \left(\frac{1}{\lambda_j}\right)^t y_j \qquad \text{for all} \quad t \in \mathbb{N}.$$

From (2.70) we therefore infer that

$$|\lambda_j| > 1 \qquad \text{for all} \quad j \in \{1, \ldots, n\} \quad \text{with} \quad \alpha_j \neq 0.$$

This is a contradiction to

$$|\lambda_j| \leq 1 \qquad \text{for all} \quad j = 1, \ldots, n.$$

Hence the assumption $R_\infty \neq \mathbb{R}^n$ is false. □

Remark. *If we define*

$$\overline{Y} = (y_1 \mid y_2 \mid \ldots \mid y_n) \qquad \textit{and} \qquad \wedge = \begin{pmatrix} \lambda_1 & & 0 \\ & \ddots & \\ 0 & & \lambda_n \end{pmatrix},$$

then it follows that

$$A^T \overline{Y} = \overline{Y} \wedge$$

which implies

$$\overline{Y}^{-1} A^T \overline{Y} = \wedge$$

and in turn

$$\overline{Y}^T A \left(\overline{Y}^T\right)^{-1} = \wedge \qquad \left(\textit{since } \left(A^{-1}\right)^T = \left(A^T\right)^{-1}\right)$$

from which

$$A \left(\overline{Y}^T\right)^{-1} = \left(\overline{Y}^T\right)^{-1} \wedge$$

follows.

This implies that A has the same eigenvalues as A^T (which holds for arbitrary matrices) and the eigenvectors of A are the column vectors of $\left(\overline{Y}^T\right)^{-1}$. Therefore A^T in Theorem 2.19 could be replaced by A. For the following let us assume that $\Omega = \mathbb{R}^m$.

For every $N \in \mathbb{N}$ let us define

$$\overline{Y}(N) = \left(B \mid AB \mid \ldots \mid A^{N-1}B\right).$$

Since U (2.53) consists of all functions $u : \mathbb{N}_0 \to \mathbb{R}^m$, it follows, for every $N \in \mathbb{N}$, that

$$E(N) = \left\{ x = \sum_{t=1}^{N} A^{N-t} B\, u(t-1) \,\middle|\, u : \mathbb{N}_0 \to \mathbb{R}^m \right\}.$$

Further we can prove

Proposition 2.20. *The following statements are equivalent:*

(i) $\operatorname{rank} \overline{Y}(N) = \operatorname{rank} \overline{Y}(N+1)$,
(ii) $E(N) = E(N+1)$,
(iii) $\left(A^N B\right) \mathbb{R}^m \subseteq E(N)$,
(iv) $\operatorname{rank} \overline{Y}(N) = \operatorname{rank} \overline{Y}(N+j)$ *for all* $j \geq 1$.

Proof.
(i) $\Rightarrow$ (ii): This is a consequence of the fact that $E(N) \subseteq E(N+1)$.
(ii) $\Rightarrow$ (iii): This follows from $\overline{Y}(N+1) = \left(\overline{Y}(N) \mid A^N B\right)$.
(iii) $\Rightarrow$ (i): $\overline{Y}(N+1) = \left(\overline{Y}(N) \mid A^N B\right)$ shows that (iii) $\Rightarrow$ (ii) and obviously we have (ii) $\Rightarrow$ (i).
(i) $\Rightarrow$ (iv): Since (i) implies (iii), it follows that

$$\left(A^{N+1} B\right) \mathbb{R}^m \subseteq A\, E(N) \subseteq E(N+1)$$

which implies $E(N+1) = E(N+2)$ and hence

$$\operatorname{rank}\left(\overline{Y}(N+1)\right) = \operatorname{rank}\left(\overline{Y}(N+2)\right).$$

(iv) $\Rightarrow$ (i): is obvious. This completes the proof. □

Now let r be the smallest integer such that $I, A, \ldots, A^{r-1}$ are linearly independent in $\mathbb{R}^{n \cdot n}$ and hence there are numbers $\alpha_{r-1}, \alpha_{r-2}, \ldots, \alpha_0 \in \mathbb{R}$ such that

$$A^r + \alpha_{r-1} A^{r-1} + \ldots + \alpha_0 I = 0.$$

Defining

$$\Phi_0(\lambda) = \lambda^r + \alpha_{r-1} \lambda^{r-1} + \ldots + \alpha_0,$$

we have $\Phi_0(A) = 0$. This monic polynomial (leading coefficient 1) is the monic polynomial of least degree for which $\Phi_0(A) = 0$ and is called the minimal polynomial of A.

The polynomial

$$\Phi(\lambda) = \det(\lambda I - A) \qquad \text{with degree} \quad n$$

is called characteristic polynomial of A, and the Cayley-Hamilton Theorem states that $\Phi(A) = 0$ which implies $r \leq n$.

This leads to

Proposition 2.21. *Let s be the degree of the minimal polynomial of A ($s \leq n$). Then there is an integer $k \leq s$ such that*

$$\operatorname{rank} \overline{Y}(1) < \operatorname{rank} \overline{Y}(2) < \ldots < \operatorname{rank} \overline{Y}(k) = \operatorname{rank} \overline{Y}(k+j) \qquad \textit{for all} \quad j \in \mathbb{N}.$$

Proof. Proposition 2.20 implies the existence of such an integer k, since $\operatorname{rank} \overline{Y}(N) \leq n$ for all $N \in \mathbb{N}$. We have to show that $k \leq s$. Let $\psi(\lambda) = \lambda^s + \alpha_{s-1}\lambda^{s-1} + \ldots + \alpha_0$ be the minimal polynomial of A. Then $\psi(A)\,B = 0$ and $A^s B\,\mathbb{R}^m \subseteq E(s)$ which implies (by Proposition 2.20) that $\operatorname{rank} \overline{Y}(s) = \operatorname{rank} \overline{Y}(s+j)$ for all $j \in \mathbb{N}$, hence $k \leq s$. □

As a consequence of Proposition 2.21 we obtain

Proposition 2.22. *If the Kalman condition (2.67) is satisfied for some $N_0 \in \mathbb{N}$, then necessarily $N_0 \leq n$ and*

$$\operatorname{rank}\left(B \mid AB \mid \ldots \mid A^{n-1}B\right) = n,$$

hence $E(n) = \mathbb{R}^n$. Conversely, if $E(n) = \mathbb{R}^n$, then Kalman's condition (2.67) is satisfied for all $N \geq n$ and

$$E(N) = \mathbb{R}^n \qquad \textit{for all} \quad N \geq n.$$

Proposition 2.21 also implies that, if $\operatorname{rank} \overline{Y}(n) < n$, then $\operatorname{rank} \overline{Y}(N) < n$ for all $N \geq n$ and

$$E(N) = E(n) \neq \mathbb{R}^n \qquad \text{for all} \quad N \geq n.$$

If we define, for every $N \in \mathbb{N}$, the $n \times n$–matrix

$$W(N) = \overline{Y}(N)\overline{Y}(N)^T = \sum_{j=0}^{N-1} A^j B B^T \left(A^j\right)^T,$$

then it follows that

$$W(N)\,\mathbb{R}^n \subseteq E(N) \qquad \text{for every} \quad N \in \mathbb{N}.$$

Now let us assume that $\operatorname{rank} \overline{Y}(N) = n$ which is equivalent to $E(N) = \mathbb{R}^n$.

Then $W(N)$ is non-singular and

$$W(N)\,\mathbb{R}^n = \mathbb{R}^n = E(N).$$

Let $y^* \in \mathbb{R}^n$ be the unique solution of

$$W(N)\,y^* = -A^N\,x_0.$$

Further let $u \in \mathbb{R}^{m\cdot N}$ be any solution of

$$\overline{Y}(N)\,u = -A^N\,x_0 \qquad \text{(see (2.65))}.$$

If we put

$$u^* = \overline{Y}(N)^T\,y^* \qquad \left(\in \mathbb{R}^{m\cdot N}\right),$$

then

$$\overline{Y}(N)^T\,u^* = -A^N\,x_0$$

and

$$\begin{aligned} ||u^*||_2^2 &= u^{*T}u^* = y^{*T}\,\overline{Y}(N)\,u^* = y^{*T}W(N)y^* \\ &= -y^{*T}A^N\,x_0 = y^*\,\overline{Y}(N)\,u = u^{*T}u \le ||u^*||_2\,||u||_2 \end{aligned}$$

which implies $||u^*||_2 \le ||u||_2$.

2.2.3 A Method for Solving the Problem of Null-Controllability

Let us equip $\mathbb{R}^m$ with the Euclidian norm $||\cdot||_2$ and consider the following

Problem (P) *For a given $N \in \mathbb{N}$ find $u : \{0,\ldots,N-1\} \to \mathbb{R}^m$ such that*

$$\sum_{t=1}^{N} A^{N-t}B\,u(t-1) = -A^N\,x_0 \tag{2.71}$$

and

$$\varphi_N(u) = \max_{t=1,\ldots,N} ||u(t-1)||_2$$

is as small as possible.

If the Kalman condition (2.67) is satisfied for some $N_0 \in \mathbb{N}$ and if $N \ge N_0$, then Problem (P) has a solution $u_N \in \mathbb{R}^{m\cdot N}$. If $\varphi_N(u_N) \le \gamma$ (see (2.68)), then we obtain solution $u_N : \mathbb{N}_0 \to \mathbb{R}^m$ of the problem of null-controllability if we define

$$u_N(t) = \Theta_m \qquad \text{for all} \quad t \ge N.$$

If $\varphi_N(u_N) > \gamma$, then N must be increased.

If the matrix A is non-singular, then (2.65) is equivalent to

$$\sum_{t=1}^{N} A^{-t} B\, u(t-1) = -x_0$$

which implies

$$\varphi_{N+1}\left(u_{N+1}\right) \leq \varphi_N\left(u_N\right) \qquad \text{for all} \quad N \geq N_0.$$

Under the assumptions of Theorem 2.19 there exists, for every $\varepsilon > 0$, some $N(\varepsilon) \in \mathbb{N}$ such that

$$\varphi_{N(\varepsilon)}\left(u_{N(\varepsilon)}\right) \leq \varepsilon$$

which implies

$$\lim_{N\to\infty} \varphi_N\left(u_N\right) = 0.$$

So we can be sure, for every choice of $\gamma > 0$, to find a solution $U_{N(\gamma)} \in \mathbb{R}^{m\cdot N(\gamma)}$ of Problem (P) with $\varphi_{N(\gamma)}\left(u_{N(\gamma)}\right) \leq \gamma$ which leads to a solution $u_{N(\gamma)} : \mathbb{N}_0 \to \mathbb{R}^m$ of the problem of null-controllability, if we define

$$u_{N(\gamma)}(t) = \Theta_m \qquad \text{for all} \quad t \geq N(\gamma).$$

In order to solve Problem (P) we replace it by

Problem (D) *Minimize*

$$\chi(y) = \sum_{k=1}^{N} \left\| B^T \left(A^{N-k}\right)^T y \right\|_2, \qquad y \in \mathbb{R}^n,$$

subject to

$$c^T y = -x_0^T \left(A^T\right)^N y = 1. \tag{2.72}$$

Let $u : \{0, \ldots, N-1\} \to \mathbb{R}^m$ be a solution of (2.65) and let $y \in \mathbb{R}^n$ satisfy (2.72). Then it follows that

$$\sum_{k=1}^{N} y^T A^{N-k} B\, u(k-1) = y^T c = 1$$

which implies

$$\max_{k=1,\ldots,N} ||u(k-1)||_2 \geq \frac{1}{\chi(y)}.$$

Now let $\widehat{y} \in \mathbb{R}^n$ be a solution of Problem (D). Then there is a multiplier $\lambda \in \mathbb{R}$ such that

$$\nabla\chi\left(\widehat{y}\right) = \lambda c. \tag{2.73}$$

This yields

$$\nabla\chi(\widehat{y}) = \sum_{k\in I(\widehat{y})} \frac{1}{||B^T\left(A^{N-k}\right)^T \widehat{y}||_2} A^{N-k}BB^T\left(A^{N-k}\right)^T \widehat{y}$$

with

$$I(\widehat{y}) = \left\{k \;\middle|\; \left\|B^T\left(A^{N-k}\right)^T \widehat{y}\right\|_2 > 0\right\}.$$

This implies

$$\lambda = \widehat{y}^T \nabla\chi(\widehat{y}) = x(\widehat{y}).$$

If we define

$$u_N(k-1) = \begin{cases} \dfrac{1}{x(\widehat{y})} \dfrac{B^T\left(A^{N-k}\right)^T \widehat{y}}{||B^T\left(A^{N-k}\right)^T \widehat{y}||_2} & \text{if} \quad k \in I(\widehat{y}), \\ \Theta_m & \text{else}, \quad k = 1,\dots,N, \end{cases} \tag{2.74}$$

then it follows that

$$\sum_{k=1}^{N} A^{N-k}B\,u_N(k-1) = c$$

and

$$||u_N(k-1)||_2 = \frac{1}{\chi(\widehat{y})} \qquad \text{for all} \quad k \in I(\widehat{y})$$

which implies

$$\max_{k=1,\dots,N} ||u_N(k-1)||_2 = \frac{1}{\chi(\widehat{y})}.$$

Hence $u : \{0,\dots,N-1\} \to \mathbb{R}^m$ solves Problem (P). This result is summarized as

Theorem 2.23. *If $\widehat{y} \in \mathbb{R}^n$ solves Problem (D), then $u : \{0,\dots,N-1\} \to \mathbb{R}^m$ defined by (2.74) solves Problem (P).*

In order to solve Problem (D) we apply the well known gradient projection method which is based on the following iteration step: Let $y^* \in \mathbb{R}^n$ with $c^T y^* = 1$ be given. (At the beginning we take $y^* = c/||c||_2^2$.) Then we calculate

$$h = \left(\frac{1}{||c||_2^2} c^T \nabla\chi(y^*)\right) c - \nabla\chi(y^*)$$

and see that

$$ch = 0$$

and

$$\nabla\chi(y^*)^T h = \frac{1}{||c||_2^2}\left(c^T \nabla\chi(y^*)\right)^2 - ||\nabla\chi(y^*)||_2^2 \le 0.$$

If $\nabla\chi(y^*)^T h = 0$, then there exists some $\lambda \geq 0$ such that (2.73) holds true which is equivalent to y^* being optimal. If $\nabla\chi(y^*)^T h < 0$, then h is a feasible direction of descent. If we determine $\widehat{\lambda} > 0$ such that

$$\chi\left(y^* + \widehat{\lambda}\,h\right) = \min_{\lambda>0} \chi\left(y^* + \lambda\,h\right), \tag{2.75}$$

then

$$\chi\left(y^* + \widehat{\lambda}\,h\right) < \chi(y^*)$$

and $c^T\left(y^* + \widehat{\lambda}\,h\right) = 1$. The next step is then performed with $y^* + \widehat{\lambda}\,h$ instead of y^*. A necessary and sufficient condition for $\widehat{\lambda} > 0$ to satisfy (2.75) is

$$\frac{\mathrm{d}}{\mathrm{d}\lambda}\chi\left(y^* + \widehat{\lambda}\,h\right) = \nabla\chi\left(y^* + \widehat{\lambda}\,h\right)^T h = 0$$

which is equivalent to

$$\sum_{k\in I\left(y^*+\widehat{\lambda}\,h\right)} \frac{h^T A^{N-k} B B^T \left(A^{N-k}\right)^T \left(y^* + \widehat{\lambda}\,h\right)}{\left\| B^T \left(A^{N-k}\right)^T \left(y^* + \widehat{\lambda}\,h\right)\right\|_2} = 0$$

and in turn to the fixed point equation

$$\widehat{\lambda} = \psi\left(\widehat{\lambda}\right)$$

with

$$\psi(\lambda) = \frac{\sum\limits_{k\in I(y^*+\lambda\,h)} \frac{h^T A^{N-k} B B^T \left(A^{N-k}\right)^T y^*}{\left\|B^T (A^{N-k})^T (y^*+\lambda\,h)\right\|_2}}{\sum\limits_{k\in I(y^*+\lambda\,h)} \frac{h^T A^{N-k} B B^T (A^{N-k})^T h}{\left\|B^T (A^{N-k})^T (y^*+\lambda\,h)\right\|_2}}.$$

In order to solve this equation we apply the iteration procedure

$$\lambda_{k+1} = \psi(\lambda_k), \qquad k \in \mathbb{N}_0$$

starting with $\lambda_0 = 0$. Let us return to the problem of fixed point controllability in Section 2.2.1 and let us assume that $g : \mathbb{R}^n \times \mathbb{R}^m \to \mathbb{R}^n$ is continuously Fréchet differentiable.

Then it follows that

$$
\begin{aligned}
&G^N\left(x_0, u(0), \ldots, u(N-1)\right)-\widehat{x} \\
&\quad=G^N\left(x_0, u(0), \ldots, u(N-1)\right)-G^N\left(\widehat{x}, \Theta_m, \ldots, \Theta_m\right) \\
&\quad\approx J_{G^N}^x\left(\widehat{x}, \Theta_m, \ldots, \Theta_m\right)\left(x_0-\widehat{x}\right)+\sum_{k=1}^N J_{G^N}^{u(k-1)}\left(\widehat{x}, \Theta_m, \ldots, \Theta_m\right) u(k-1) \\
&\quad=J_g^x\left(\widehat{x}, \Theta_m\right)^N\left(x_0-\widehat{x}\right)+\sum_{k=1}^N J_g^x\left(\widehat{x}, \Theta_m\right)^{N-k} J_g^u\left(\widehat{x}, \Theta_m\right) u(k-1)
\end{aligned}
$$

where

$$J_g^x\left(\widehat{x}, \Theta_m\right)=\left(g_{i x_j}\left(\widehat{x}, \Theta_m\right)\right)_{i, j=1, \ldots, n}$$

and

$$J_g^u\left(\widehat{x}, \Theta_m\right)=\left(g_{i u_k}\left(\widehat{x}, \Theta_m\right)\right)_{\substack{i=1, \ldots, n \\ k=1, \ldots, m}} .$$

Therefore we replace equation (2.58) by

$$\sum_{k=1}^N J_g^x\left(\widehat{x}, \Theta_m\right)^{N-k} J_g^u\left(\widehat{x}, \Theta_m\right) u(k-1)=-J_g^x\left(\widehat{x}, \Theta_m\right)^N\left(x_0-\widehat{x}\right) \tag{2.76}$$

and solve the problem of finding $u:\{0, \ldots, N-1\} \rightarrow \mathbb{R}^m$ which solves (2.76) and minimizes

$$\varphi_N(u)=\max_{k=1, \ldots, N}\|u(k-1)\|_2 .$$

Such a $u:\{0, \ldots, N-1\} \rightarrow \mathbb{R}^m$ is then taken as an approximate solution of (2.58).

The above problem has the form of Problem (P) at the beginning of this section and can be solved by the method described above.

Finally we consider a special case in which the problem of fixed point controllability is reduced to a sequence of such problems which can be solved more easily.

For this purpose we consider the system

$$x(t+1)=g_0(x(t))+\sum_{j=1}^m g_j(x(t))\, u_j(t), \qquad t \in \mathbb{N}_0, \tag{2.77}$$

where $g_j: \mathbb{R}^n \rightarrow \mathbb{R}^n,\ j=1, \ldots, m$, are continuous vector functions.

For every control function $u: \mathbb{N}_0 \rightarrow \mathbb{R}^m$ there is exactly one function $x: \mathbb{N}_0 \rightarrow \mathbb{R}^m$ which satisfies (2.77) and the initial condition

$$x(0)=x_0, \qquad x_0 \in \mathbb{R}^n \text{ given.} \tag{2.78}$$

We denote it by $x = x(u)$. We assume that the uncontrolled system

$$x(t+1) = g_0(x(t)), \qquad t \in \mathbb{N},$$

has a fixed point $\widehat{x} \in \mathbb{R}^n$ which then solves the system

$$\widehat{x} = g_0(\widehat{x}).$$

We again assume that the set U of admissible control functions is given by (2.53) where $\Omega \subseteq \mathbb{R}^m$ is a subset with $\Theta_m \in \Omega$.

Let us define

$$\widetilde{g}_0(x) = g_0(x) - x, \qquad x \in \mathbb{R}^n.$$

Then (2.77) can be rewritten in the form

$$x(t+1) = x(t) + \widetilde{g}_0(x(t)) + \sum_{j=1}^{m} g_j(x(t))\, u_j(t), \qquad t \in \mathbb{N}_0. \tag{2.79}$$

In order to find some $N \in \mathbb{N}_0$ and a control function $u \in U$ with (2.55) such that the solution $x : \mathbb{N}_0 \to \mathbb{R}^n$ of (2.77) satisfies the end condition (2.56) we apply an iterative method: Starting with some $N_0 \in \mathbb{N}_0$ and some $u^0 \in U$ (for instance $u^0(t) = \Theta_m$ for all $t \in \mathbb{N}_0$) we construct a sequence $(N_k)_{k\in\mathbb{N}}$ in $\mathbb{N}_0$ and a sequence $\left(u^k\right)_{k\in\mathbb{N}}$ in U as follows:

If $N_{k-1} \in \mathbb{N}_0$ and $u^{k-1} \in U$ are determined, we calculate $x\left(u^{k-1}\right) : \mathbb{N}_0 \to \mathbb{R}^n$ as the solution of (2.78) and (2.79) for $u = u^{k-1}$. Then we determine $N_k \in \mathbb{N}_0$ and $u^k \in U$ such that

$$u^k(t) = \Theta_m \qquad \text{for all} \quad t \geq N_k \tag*{$(2.55)_k$}$$

and the solution $x\left(u^k\right) : \mathbb{N}_0 \to \mathbb{R}^n$ of (2.78) and

$$\begin{aligned} x\left(u^k\right)(t+1) &= x\left(u^k\right)(t) + \widetilde{g}_0\left(x\left(u^{k-1}\right)(t)\right) \\ &\quad + \sum_{j=1}^{m} g_j\left(x\left(u^{k-1}\right)(t)\right) u_j^{k+1}(t), \qquad t \in \mathbb{N}_0, \end{aligned} \tag*{$(2.79)_k$}$$

satisfies the end condition

$$x\left(u^k\right)(N_k) = \widehat{x}. \tag*{$(2.56)_k$}$$

If we put

$$x^k = x_0 + \sum_{t=1}^{N_k} \widetilde{g}_0\left(x\left(u^{k-1}\right)(t-1)\right)$$

and

$$B^k(t-1) = \left(g_1\left(x\left(u^{k-1}\right)(t-1)\right) \mid \ldots \mid g_m\left(x\left(u^{k-1}\right)(t-1)\right)\right),$$

then the end condition $(2.56)_k$ is equivalent to

$$\sum_{t=1}^{N} B^k(t-1)\, u^{k-1}(t-1) = \widehat{x} - x^k.$$

2.2.4 Stabilization of Controlled Systems

Let $g : \mathbb{R}^n \times \mathbb{R}^m \to \mathbb{R}^n$ be a continuous mapping and let H be a family of continuous mappings $h : \mathbb{R}^n \to \mathbb{R}^m$. If we define, for every $h \in H$, the mapping $f_h : \mathbb{R}^n \to \mathbb{R}^n$ by

$$f_h(x) = g(x, h(x)), \qquad x \in \mathbb{R}^n,$$

then f_h is continuous and $(\mathbb{R}^n, f_h)$ is a time-discrete autonomous dynamical system. Let $\widehat{x} \in \mathbb{R}^n$ be a fixed point of

$$f(x) = g\left(x, \Theta_m\right), \qquad x \in \mathbb{R}^n.$$

Further we assume that

$$h\left(\widehat{x}\right) = \Theta_m \qquad \text{for all} \quad h \in H$$

which implies that $\widehat{x}$ is a fixed point of all f_h, $h \in H$. After these preparations we can formulate the

Problem of Stabilization

Find $h \in H$ such that $\{\widehat{x}\}$ is asymptotically stable with respect to f_h.

We assume that $g : \mathbb{R}^n \times \mathbb{R}^m \to \mathbb{R}^n$ and every mapping $h \in H$ are continuously Fréchet differentiable. Then every mapping $f_h : \mathbb{R}^n \to \mathbb{R}^n$, $h \in H$, is also continuously Fréchet differentiable and, for every $x \in \mathbb{R}^n$, its Jacobi matrix is given by

$$J_{f_h}(x) = J_g^x(x, h(x)) + J_g^u(x, h(x))\, J_h^x(x)$$

where

$$J_g^x(x, h(x)) = \left(g_{i x_j}(x, h(x))\right)_{\substack{i=1,\ldots,n \\ j=1,\ldots,n}}$$

and

$$J_g^u(x, h(x)) = \left(g_{i u_k}(x, h(x))\right)_{\substack{i=1,\ldots,n \\ k=1,\ldots,m}}, \qquad J_h^x(x) = \left(h_{i x_j}(x)\right)_{\substack{i=1,\ldots,m \\ j=1,\ldots,n}}.$$

From the Corollary of Theorem 1.5 we then obtain the

Theorem 2.24.
(a) *Let the spectral radius $\varrho\left(J_{f_h}\left(\widehat{x}\right)\right) < 1$. Then $\widehat{x}$ is asymptotically stable with respect to f_h.*
(b) *Let $\left(J_{f_h}\left(\widehat{x}\right)\right)$ be invertible and let all the eigenvalues of $\varrho\left(J_{f_h}\left(\widehat{x}\right)\right)$ be larger than 1 in absolute value. Then $\widehat{x}$ is unstable with respect to f_h.*

Special cases:

(a) Let
$$g(x,u) = Ax + Bu, \qquad x \in \mathbb{R}^n,\ u \in \mathbb{R}^m,$$
where A is a real $n \times n$–matrix and B a real $n \times m$–matrix, respectively. Further let H be the family of all linear mappings $h : \mathbb{R}^n \to \mathbb{R}^m$ which are given by
$$h(x) = Cx, \qquad x \in \mathbb{R}^n,$$
where C is an arbitrary real $m \times n$–matrix, respectively.

If we choose $\widehat{x} = \Theta_n$, then
$$f(\Theta_n) = g(\Theta_n, \Theta_m) = \Theta_n$$
and
$$h(\Theta_n) = \Theta_m \qquad \text{for all} \quad h \in H.$$
Finally we have $J_h(x) = C$,
$$J_g^x(x, h(x)) = A \qquad \text{and} \qquad J_g^u(x, h(x)) = B$$
for all $x \in \mathbb{R}^n$ and $h \in H$ which implies
$$J_{f_h}(x) = A + BC \qquad \text{for all} \quad x \in \mathbb{R}^n \quad \text{and} \quad h \in H.$$
Thus $\widehat{x} = \Theta_n$ is asymptotically stable with respect to f_h, if
$$\varrho(A + BC) < 1,$$
and unstable with respect to f_h, if all the eigenvalues of $A + BC$ are larger than one in absolute value.

(b) Let
$$g(x,u) = F(x) + B(x)\,u, \qquad x \in X,\ u \in \mathbb{R}^m,$$
where $F : X \to X$, $X \subseteq \mathbb{R}^n$ open, is continuously Fréchet differentiable and $B(x) = (b_1(x), \ldots, b_m(x))$, $x \in X$, where $b_j : X \to \mathbb{R}^n$, $j = 1, \ldots, n$, are also continuously Fréchet differentiable. Let again H be the family of all linear mappings $h : \mathbb{R}^n \to \mathbb{R}^m$ which are given by
$$h(x) = Cx, \qquad x \in \mathbb{R}^n.$$
Finally, we assume that $\Theta_n \in X$ and $F(\Theta_n) = \Theta_n$.

If we choose $\widehat{x} = \Theta_n$, then
$$f(\Theta_n) = g(\Theta_n, \Theta_m) = \Theta_n$$
and
$$h(\Theta_n) = \Theta_m \qquad \text{for all} \quad h \in H.$$

Further we obtain

$$J_h^x(x) = C,$$

$$J_g^x(x, h(x)) = J_F(x) + \sum_{j=1}^{m} J_{b_j}(x)\, h_j(x) \qquad \text{and} \qquad J_g^u(x, h(x)) = B(x)$$

for all $x \in X$ and $h \in H$ which implies

$$J_{f_h}(x) = J_F(x) + \sum_{j=1}^{m} J_{b_j}(x)\, h_j(x) + B(x)C \qquad \text{for} \quad x \in X,\ h \in H,$$

hence

$$J_{f_h}(\Theta_n) = J_F(\Theta_n) + B(\Theta_n)\, C \qquad \text{for all} \quad h \in H.$$

Thus $\widehat{x} = \Theta_n$ is asymptotically stable with respect to f_h, if

$$\varrho\left(J_F(\Theta_n) + B(\Theta_n)\, C\right) < 1$$

and unstable with respect to f_h, if all the eigenvalues of $J_F(\Theta_n) + B(\Theta_n)\, C$ are larger than one in absolute value.

2.2.5 Applications

a) An Emission Reduction Model

We pick up the emission reduction model that was treated as uncontrolled system in Section 1.3.6 and as controlled system in Section 2.2.1. Here we concentrate on the controlled system which we linearize at a fixed point $\left(\widehat{E}^T, \Theta_r^T\right)^T$, $\widehat{E} \in \mathbb{R}^r$, of the uncontrolled system which leads to a linear control system of the form

$$x(t+1) = A\,x(t) + B\,u(t), \qquad t \in \mathbb{N}_0,$$

with

$$A = \begin{pmatrix} I_r & C \\ 0_r & D \end{pmatrix}, \qquad B = \begin{pmatrix} C \\ D \end{pmatrix},$$

where I_r and 0_r is the $r \times r-$unit and zero-matrix, respectively, and

$$C = \begin{pmatrix} em_{11} & \cdots & em_{1r} \\ \vdots & \ddots & \vdots \\ em_{r1} & \cdots & em_{rr} \end{pmatrix}, \qquad D = \begin{pmatrix} 1 - \lambda_1 M_1^* \widehat{E}_1 & & 0 \\ & \ddots & \\ 0 & & 1 - \lambda_r M_r^* \widehat{E}_r \end{pmatrix}.$$

This implies

$$A^k B = \begin{pmatrix} C\left(I_r + D + \ldots + D^k\right) \\ D^{k+1} \end{pmatrix} \qquad \text{for all} \quad k \in \mathbb{N}_0.$$

We consider the problem of null-controllability as being discussed in Section 2.2.2. Let us assume that C and D are non-singular. Then it follows that the matrices A and

$$\begin{pmatrix} C & C\left(I_r + D\right) \\ D & D^2 \end{pmatrix}$$

are non-singular which implies that the Kalman condition (2.67) is satisfied for $N_0 = 2$. Let $d_1, \ldots, d_r$ be the diagonal elements of D. Thus the non-singularity of D is equivalent to

$$d_i \neq 0 \qquad \text{for all} \quad i = 1, \ldots, r.$$

If all $d_i \neq 1$ for $i = 1, \ldots, r$, then it follows (see Section 1.3.6) that the eigenvectors corresponding to the eigenvalues

$$\mu_i = 1 \quad \text{for} \quad i = 1, \ldots, r \qquad \text{and} \qquad \mu_{i+r} = d_i \quad \text{for} \quad i = 1, \ldots, r$$

of A are linearly independent which also holds true for A^T (which has the same eigenvalues) (see Section 2.1.2).

If

$$|d_i| \leq 1 \qquad \text{for all} \quad i = 1, \ldots, r$$

and

$$\Omega = \{u \in \mathbb{R}^r \mid ||u|| \leq \gamma\}$$

for some $\gamma > 0$ where $||\cdot||$ is any norm in $\mathbb{R}^r$, then by Theorem 2.19 the problem of null-controllability has a solution for every choice of $x_0 = \left(x_0^{1\,T}, x_0^{2\,T}\right)^T \in \mathbb{R}^{2r}$. This problem can be solved with the aid of Problem (P) in Section 2.2.3 which reads as follows in this case: For a given $N \in \mathbb{N}$ find $u : \{0, \ldots, N-1\} \to \mathbb{R}^r$ such that

$$\sum_{k=1}^{N} \begin{pmatrix} C\left(I_r + D + \ldots + D^k\right) \\ D^{k+1} \end{pmatrix} u(k-1)$$
$$= -\begin{pmatrix} I_r & C\left(I_r + D + \ldots + D^{N-1}\right) \\ 0_r & D^N \end{pmatrix} \begin{pmatrix} x_0^1 \\ x_0^2 \end{pmatrix}$$

and

$$\varphi_N(u) = \max_{k=1,\ldots,N} ||u(k-1)||_2$$

is minimized (where $||\cdot||_2$ denotes the Euclidean norm in $\mathbb{R}^r$).

Finally we illustrate the method by two numerical examples.

Let $m = 3$. In both cases we choose

$$x_0^{1\,T} = x_0^{2\,T} = (1, 1, 1). \tag{2.80}$$

At first we choose

$$C = \begin{pmatrix} 0.8 & 0.5 & -0.5 \\ 0.2 & 0.2 & 0.3 \\ 0.4 & 0.3 & 0.2 \end{pmatrix} \quad \text{and} \quad D = \begin{pmatrix} 0.1 & 0 & 0 \\ 0 & -0.2 & 0 \\ 0 & 0 & 0.1 \end{pmatrix}$$

and obtain Fig. 2.1:

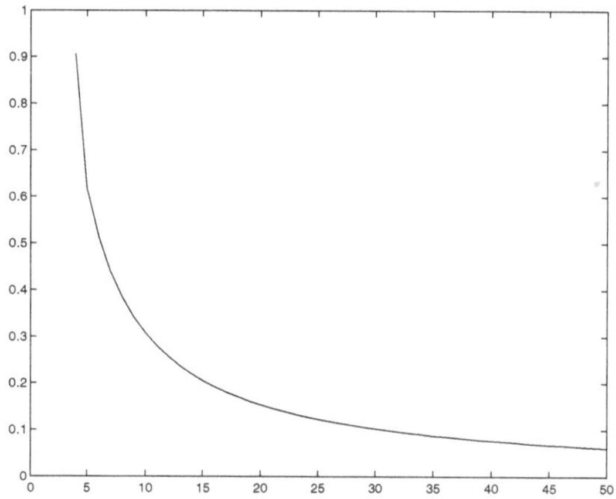

Fig. 2.1.
Ordinate, $\varphi_N(u_N) = \frac{1}{\chi(\hat{y}_N)}$
Abscissa, N

Next we choose

$$C = \begin{pmatrix} 0.8 & 0.5 & 0.5 \\ 0.2 & 0.2 & 0.3 \\ 0.4 & 0.1 & 0.2 \end{pmatrix} \quad \text{and} \quad D = \begin{pmatrix} 0.2 & 0 & 0 \\ 0 & 0.8 & 0 \\ 0 & 0 & 0.6 \end{pmatrix}$$

and get Fig. 2.2.

b) A Controlled Predator-Prey Model

We pick up the predator prey model that has been discussed in Section 1.3.7 a) whose controlled version we assume to be of the form

$$\begin{aligned} x_1(t+1) &= x_1(t) + a\,x_1(t) - b\,x_1(t)x_2(t) - x_1(t)\,u_1(t), \\ x_2(t+1) &= x_2(t) - c\,x_2(t) + d\,x_1(t)x_2(t) - x_2(t)\,u_2(t) \qquad t \in \mathbb{N}_0, \end{aligned}$$

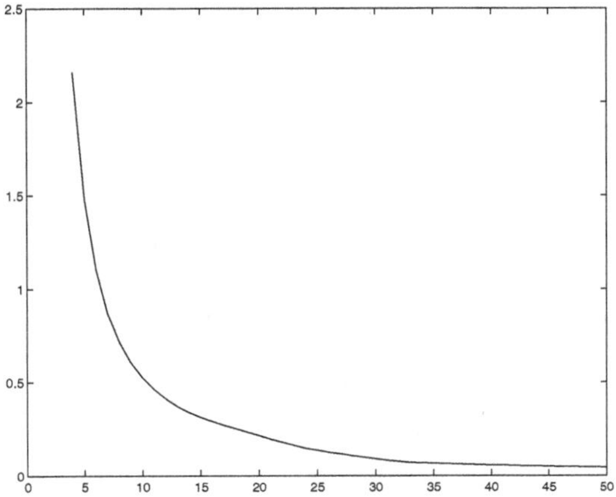

Fig. 2.2.
Ordinate, $\varphi_N(u_N) = \frac{1}{\chi(\hat{y}_N)}$
Abscissa, N

where $a > 0,\ 0 < c < 1,\ b > 0,\ d > 0,\ x_1(t)$ and $x_2(t)$ denote the density of the prey and predator population at time t, respectively, and $u_1, u_2 : \mathbb{N}_0 \to \mathbb{R}$ are control functions.

If we define

$$\widetilde{g}_0\left(x_1(t), x_2(t)\right) = \begin{pmatrix} a\,x_1(t) - b\,x_1(t)x_2(t) \\ -c\,x_1(t) + d\,x_1(t)x_2(t) \end{pmatrix}$$

and

$$g_1\left(x_1(t), x_2(t)\right) = \begin{pmatrix} -x_1(t) \\ 0 \end{pmatrix}, \qquad g_2\left(x_1(t), x_2(t)\right) = \begin{pmatrix} 0 \\ -x_2(t) \end{pmatrix},$$

then the system can be rewritten in the form

$$x(t+1) = x(t) + \widetilde{g}_0(x(t)) + \sum_{j=1}^{2} g_j(x(t))\, u_j(t), \qquad t \in \mathbb{N}_0, \tag{2.81}$$

with $x(t) = (x_1(t), x_2(t))^T$. This is exactly the system (2.79) for $m = 2$.

In addition we assume an initial condition

$$x(0) = \begin{pmatrix} x_{01} \\ x_{02} \end{pmatrix} = x_0. \tag{2.82}$$

The uncontrolled system

$$x(t+1) = x(t) + \widetilde{g}_0(x(t)), \qquad t \in \mathbb{N}_0,$$

has $\widehat{x} = \left(\frac{c}{d}, \frac{a}{b}\right)^T$ as fixed point.

We assume the set U of admissible control functions to be given by

$$U = \left\{u : \mathbb{N}_0 \to \mathbb{R}^2 \mid ||u(t)||_2 \leq \gamma \text{ for all } t \in \mathbb{N}_0\right\}$$

where $\gamma > 0$ is a given constant and $||\cdot||_2$ is the Euclidean norm. For every $u \in U$ we denote the unique solution $x : \mathbb{N}_0 \to \mathbb{R}^2$ of (2.81) and (2.82) by $x(u)$. Our aim now consists of finding some $N \in \mathbb{N}_0$ and a control function $u \in U$ with

$$u(t) = \Theta_2 \qquad \text{for all} \quad t \geq N \tag{2.83}$$

such that the solution $x : \mathbb{N}_0 \to \mathbb{R}^2$ of (2.81), (2.82) satisfies the end condition

$$x(N) = \widehat{x}. \tag{2.84}$$

In order to find a solution of this problem we apply the iteration method described in Section 2.2.3. In the $k-$th step of this procedure we have, for a given $u^{k-1} \in U$, to find some $N_k \in \mathbb{N}_0$ and a $u^k \in U$ with

$$u^k(t) = \Theta_2 \qquad \text{for} \quad t \geq N_k$$

such that

$$\sum_{t=1}^{N_k} \begin{pmatrix} x_1\left(u^{k-1}\right)(t-1) & 0 \\ 0 & x_2\left(u^{k-1}\right)(t-1) \end{pmatrix} \begin{pmatrix} u_1^k(t-1) \\ u_2^k(t-1) \end{pmatrix} = x^k - \widehat{x},$$

where

$$x^k = x_0 + \sum_{t=1}^{N_k} \widetilde{g}_0\left(x\left(u^{k-1}\right)(t-1)\right),$$

and

$$||u^k(t-1)||_2 \leq \gamma \qquad \text{for} \quad t = 1, \ldots, N_k.$$

For this we can apply the method developed in Section 2.2.3.

c) Control of a Planar Pendulum with Moving Suspension Point

We consider a non-linear planar pendulum of length $l(>0)$ whose movement is controlled by moving its suspension point with acceleration $u = u(t)$ along a horizontal straight line.

If we denote the deviation angle from the orthogonal position of the pendulum by $\varphi = \varphi(t)$, then the movement of the pendulum is governed by the differential equation

$$\ddot{\varphi}(t) = -\frac{g}{l}\sin\varphi(t) - \frac{u(t)}{l}\cos\varphi(t), \qquad t \in \mathbb{R}, \tag{2.85}$$

where g denotes the gravity constant.

For $t = 0$ initial conditions are given by

$$\varphi(0) = \varphi_0 \qquad \text{and} \qquad \dot{\varphi}(0) = \dot{\varphi}_0.$$

Now we discretize the differential equation by introducing a time step length $h > 0$ and replacing the second derivative $\ddot{\varphi}(t)$ by $\frac{1}{h}(\varphi(t+2h) - 2\varphi(t+h) + \varphi(t))$ thus obtaining the difference equation

$$\varphi(t+2h) = 2\varphi(t+h) - \varphi(t) - \frac{gh^2}{l}\sin\varphi(t) - \frac{u(t)h^2}{l}\cos\varphi(t), \qquad t \in \mathbb{R}.$$

If we define

$$y_1(t) = \varphi(t) \qquad \text{and} \qquad y_2(t) = \varphi(t+h)$$

then we obtain

$$\begin{aligned} y_1(t+h) &= y_2(t), \\ y_2(t+h) &= 2y_2(t) - y_1(t) - \frac{gh^2}{l}\sin y_1(t) - \frac{u(t)h^2}{l}\cos y_1(t), \qquad t \in \mathbb{R}. \end{aligned}$$

Finally we define functions $x_1 : \mathbb{N}_0 \to \mathbb{R}$ and $x_2 : \mathbb{N}_0 \to \mathbb{R}$ by putting

$$x_1(n) = y_1(n \cdot h) \quad \text{and} \quad x_2(n) = y_2(n \cdot h), \qquad n \in \mathbb{N}_0,$$

and obtain the system

$$\begin{aligned} x_1(t+1) &= x_2(t), \\ x_2(t+1) &= 2x_2(t) - x_1(t) - \frac{gh^2}{l}\sin x_1(t) - \frac{u(t)h^2}{l}\cos x_1(t), \end{aligned} \tag{2.49'}$$

$t \in \mathbb{N}_0$ which is of the form (2.79) for $m = 1$ with

$$\widetilde{g}_0(x_1(t), x_2(t)) = \begin{pmatrix} x_2(t) - x_1(t) \\ x_2(t) - x_1(t) - \frac{gh^2}{l}\sin x_1(t) \end{pmatrix},$$

$$g_1(x_1(t), x_2(t)) = \begin{pmatrix} 0 \\ -\frac{h^2}{l}\cos x_1(t) \end{pmatrix}, \qquad t \in \mathbb{N}_0.$$

In addition we assume initial conditions

$$x_1(0) = \varphi_0 \qquad \text{and} \qquad x_2(0) = \dot{\varphi}_0. \tag{2.50'}$$

The uncontrolled system

$$\begin{aligned} x_1(t+1) &= x_2(t), \\ x_2(t+1) &= 2x_2(t) - x_1(t) - \frac{gh^2}{l} \sin x_1(t) \end{aligned}$$

has $\widehat{x} = (0,0)$ as fixed point.

We assume the set U of admissible control functions to be given by

$$U = \{u : \mathbb{N}_0 \to \mathbb{R} \mid |u(t)| \leq \gamma \text{ for all } t \in \mathbb{N}_0\}$$

where $\gamma > 0$ is a given constant.

For every $u \in U$ we denote the unique solution $x : \mathbb{N}_0 \to \mathbb{R}^2$ of (2.49') and (2.50') by $x(u)$.

Our aim consists of finding some $N \in \mathbb{N}_0$ and a control function $u \in U$ with

$$u(t) = 0 \qquad \text{for all} \quad t \geq N \tag{2.55'}$$

such that the solution $x : \mathbb{N}_0 \to \mathbb{R}^2$ of (2.49'), (2.50') satisfies the end condition

$$x(N) = \Theta_2.$$

This condition is equivalent to

$$x_2(N) = x_2(N-1) = 0.$$

The $k-$th step of the iteration method described in Section 2.2.3 for the solution of this problem then reads as follows: Let $u^{k-1} \in U$ be given. Then we determine $N_k \in \mathbb{N}_0$ and $u^k \in U$ such that

$$u^k(t) = 0 \qquad \text{for all} \quad t \geq N_k$$

and the solution $x\left(u^k\right) : \mathbb{N}_0 \to \mathbb{R}^n$ of (2.50') and

$$\begin{aligned} x_1\left(u^1(t+1)\right) &= x_1\left(u^k\right)(t) + x_2\left(u^{k-1}\right)(t) - x_1\left(u^{k-1}\right)(t), \\ x_2\left(u^1(t+1)\right) &= x_2\left(u^k\right)(t) + x_2\left(u^{k-1}\right)(t) - x_1\left(u^{k-1}\right)(t) \\ &\quad - \frac{gh^2}{l} \sin x_1\left(u^{k-1}\right)(t) - \frac{h^2}{l} \cos x_1\left(u^{k-1}\right)(t)\, u^k(t), \end{aligned}$$

$t \in \mathbb{N}_0$, satisfies the end conditions

$$x_2\left(u^k\right)(N_k) = x_2\left(u^k\right)(N_k - 1) = 0.$$

These are equivalent to

$$
\begin{aligned}
-\frac{1}{l}\sum_{t=1}^{N_k-1} & \cos x_1\left(u^{k-1}\right)(t-1)\, u^k(t-1) \\
&= -\dot\varphi_0 - x_2\left(u^{k-1}\right)(N_k-2) + \varphi_0 - \frac{g}{l}\sum_{t=1}^{N_k-1} \sin x_1\left(u^{k-1}\right)(t-1) \\
& \quad -\frac{1}{l}\cos x_1\left(u^{k-1}\right)(N_k-1)\, u^k\,(N_k-1) \\
&= -x_2\left(u^{k-1}\right)(N_k-1) + x_2\left(u^{k-1}\right)(N_k-2) \\
& \quad -\frac{g}{l}\sin x_1\left(u^{k-1}\right)(N_k-1)\,.
\end{aligned}
$$

Let us consider the special case $N_k = 2$. Then it follows that

$$
\begin{aligned}
-\frac{1}{l}\cos\varphi_0\, u^k(0) &= \varphi_0 - \frac{g}{l}\sin\varphi_0, \\
-\frac{1}{l}\cos\varphi_0\, u^k(1) &= -\dot\varphi_0 + \varphi_0 + \frac{gh^2}{l}\sin\varphi_0 - \frac{g}{l}\sin\dot\varphi_0 - \frac{h^2}{l}\cos\dot\varphi_0\, u^{k-1}(0).
\end{aligned}
$$

Let us assume that

$$
\cos\varphi_0 \neq 0 \qquad \text{and} \qquad \cos\dot\varphi_0 \neq 0.
$$

Then it follows for all $k \geq 1$ that

$$
\begin{aligned}
u^k(0) &= -\frac{l\varphi_0}{\cos\varphi_0} + g\tan\varphi_0, \\
u^k(1) &= \frac{l\left(\varphi_0 - \dot\varphi_0\right)}{\cos\dot\varphi_0} - gh^2\frac{\sin\varphi_0}{\cos\dot\varphi_0} + g\tan\dot\varphi_0 + h^2 u^{k-1}(0).
\end{aligned}
$$

2.3 The Time-Discrete Non-Autonomous Case

2.3.1 The Problem of Fixed Point Controllability

We consider a system of difference equations of the form

$$x(t+1) = g_t(x(t), u(t)), \qquad t \in \mathbb{N}_0, \tag{2.86}$$

where $(g_t)_{t \in \mathbb{N}_0}$ is a sequence of continuous vector functions $g_t : \mathbb{R}^n \times \mathbb{R}^m \to \mathbb{R}^n$, $u : \mathbb{N}_0 \to \mathbb{R}^m$ is a given vector function which is called a control function. The vector function $x : \mathbb{N}_0 \to \mathbb{R}^n$ which is called a state function is uniquely defined by (2.86), if we require an initial condition

$$x(0) = x_0 \tag{2.87}$$

for some given vector $x_0 \in \mathbb{R}^n$.

If we fix the control function $u : \mathbb{N}_0 \to \mathbb{R}^m$ and define

$$f_t(x) = g_t(x, u(t)), \qquad x \in \mathbb{R}^n,\ t \in \mathbb{N}_0, \tag{2.88}$$

then, for every $t \in \mathbb{N}_0$, $f_t : \mathbb{R}^n \to \mathbb{R}^n$ is a continuous mapping and $\left(\mathbb{R}^n, (f_t)_{t \in \mathbb{N}_0}\right)$ is a non-autonomous time-discrete dynamical system which is controlled by the function $u : \mathbb{N}_0 \to \mathbb{R}^m$. If

$$u(t) = \Theta_m \qquad \text{for all} \quad t \in \mathbb{N}_0,$$

then the system (2.88) is called uncontrolled. Let us assume that the uncontrolled system (2.88) admits a fixed point $\widehat{x} \in \mathbb{R}^n$ which then solves the equations

$$\widehat{x} = g_t\left(\widehat{x}, \Theta_m\right), \qquad \text{for all} \quad t \in \mathbb{N}. \tag{2.89}$$

Now let $\Omega \subseteq \mathbb{R}^m$ be a subset with $\Theta_m \in \Omega$. Then we define the set of admissible control functions by

$$U = \left\{u : \mathbb{N}_0 \to \mathbb{R}^m \mid u(t) \in \Omega \text{ for all } t \in \mathbb{N}_0\right\}. \tag{2.90}$$

After these preparations we can formulate the

Problem of Local Fixed Point Controllability

Given a fixed point $\widehat{x} \in \mathbb{R}^n$ of the system

$$x(t+1) = g_t\left(x(t), \Theta_m\right), \qquad t \in \mathbb{N}_0, \tag{2.91}$$

i.e., a solution $\widehat{x}$ of the equations (2.89) and an initial state $x_0 \in \mathbb{R}^n$, find some $N \in \mathbb{N}_0$ and a control function $u \in U$ with

$$u(t) = \Theta_m, \qquad \text{for all} \quad t \geq N \tag{2.92}$$

such that the solution $x : \mathbb{N}_0 \to \mathbb{R}^n$ of (2.86), (2.87) satisfies the end condition

$$x(N) = \widehat{x} \tag{2.93}$$

(which implies $x(t) = \widehat{x}$ for all $t \geq N$). Let us assume that, for every $t \in \mathbb{N}_0$, the Jacobi matrices

$$A_t = \frac{\partial g_t}{\partial x}\left(\widehat{x}, \Theta_m\right) \qquad \text{and} \qquad B_t = \frac{\partial g_t}{\partial x}\left(\widehat{x}, \Theta_m\right)$$

exist. Then it follows that, for every $t \in \mathbb{N}_0$,

$$\begin{aligned} x(t+1) - \widehat{x} &= g_t(x(t), u(t)) - g_t\left(\widehat{x}, \Theta_m\right) \\ &\approx A_t\,\left(x(t) - \widehat{x}\right) + B_t\,u(t). \end{aligned}$$

Therefore we replace the system (2.86) by

$$h(t+1) = A_t\,h(t) + B_t\,u(t), \qquad t \in \mathbb{N}_0, \tag{2.94}$$

and the initial condition (2.87) by

$$h(0) = x_0 - \widehat{x}. \tag{2.95}$$

The end condition (2.93) is replaced by

$$h(N) = \Theta_n = \text{ zero vector of } \mathbb{R}^n. \tag{2.96}$$

Now we consider the

Problem of Local Controllability

Find $N \in \mathbb{N}_0$ and $u \in U$ with

$$u(t) = \Theta_m \qquad \text{for all} \quad t \geq N$$

such that the corresponding solution $h : \mathbb{N}_0 \to \mathbb{R}^n$ of (2.94), (2.95) satisfies the end condition (2.96) which implies

$$h(t) = \Theta_n \qquad \text{for all} \quad t \geq N.$$

From (2.94) and (2.95) we conclude that, for every $N \in \mathbb{N}$,

$$h(N) = A_{N-1} \cdots A_0\,\left(x_0 - \widehat{x}\right) + \sum_{k=1}^{N} A_{N-1} \cdots A_k B_{k-1}\,u(k-1),$$

if, for $k = N$, we put $A_{N-1} \cdots A_k = I = n \times n-$unit matrix.

Therefore the end condition (2.96) is equivalent to

$$\sum_{k=1}^{N} A_{N-1}A_{N-2}\cdots A_k B_{k-1}\, u(k-1) = -A_{N-1}A_{N-2}\cdots A_0\left(x_0 - \widehat{x}\right). \quad (2.97)$$

Let us assume that, for some $N_0 \in \mathbb{N}$,

$$\operatorname{rank}\left(B_{N_0-1} \mid A_{N_0-1}B_{N_0-2} \mid \ldots \mid A_{N_0-1}\cdots A_1 B_0\right) = n. \quad (2.98)$$

Then, for $N = N_0$, the system (2.97) has a solution $\left(u(0)^T, u(1)^T, \ldots, u\left(N_0-1\right)^T\right)^T \in \mathbb{R}^{m\cdot N_0}$ where $x_0 \in \mathbb{R}^N$ can be chosen arbitrarily. If we define

$$u(t) = \Theta_m \qquad \text{for all} \quad t \geq N_0, \quad (2.99)$$

then we obtain a control function $u : \mathbb{N}_0 \to \mathbb{R}^m$ such that the corresponding solution $h : \mathbb{N}_0 \to \mathbb{R}^n$ of (2.94) and (2.95) satisfies the end condition (2.96) for $N = N_0$.

If A_k is non-singular for all $k \in \mathbb{N}_0$, then the assumption (2.97) implies that (2.98) holds true for all $N \geq N_0$ instead of N_0.

For instance, if we replace N_0 by $N_0 + 1$, then

$$\begin{aligned}&\left(B_{N_0} \mid A_{N_0}B_{N_0-1} \mid A_{N_0}A_{N_0-1}B_{N_0-2} \mid \ldots \mid A_{N_0}\cdots A_1 B_0\right)\\ &= \underbrace{\left(B_{N_0} \mid \underbrace{A_{N_0}}_{\text{non-singular}} \underbrace{\left(B_{N_0-1} \mid A_{N_0-1}B_{N_0-2} \mid \ldots \mid A_{N_0-1}\cdots A_1 B_0\right)}_{\text{rank}\,=\,n}\right)}_{\Longrightarrow\ \text{rank}\,=\,n}.\end{aligned}$$

So the system (2.97) has a solution $\left(u(0)^T, u(1)^T, \ldots, u\left(N_0-1\right)^T\right)^T$ for all $N \geq N_0$.

Next we assume that $\Omega \subseteq \mathbb{R}^m$ is convex, has Θ_m as interior point and satisfies $u \in \Omega \;\Rightarrow\; -u \in \Omega$. Let us define, for every $N \in \mathbb{N}$, the set

$$R(N) = \left\{ x = \sum_{k=1}^{N} A_{N-1}A_{N-2}\cdots A_k B_{k-1}\, u(k-1) \,\middle|\, u \in U \right\}.$$

Then we can prove (see Theorem 2.18)

Theorem 2.25. *If, for some $N_0 \in \mathbb{N}$, condition (2.98) is satisfied and if A_k is non-singular for all $k \in \mathbb{N}_0$, then Θ_n is an interior point of $R(N)$ for all $N \geq N_0$.*

Proof. Let us assume that $\Theta_n (\in R(N))$ is not an interior point of $R(N)$ for some $N \geq N_0$. Thus $R(N)$ must be contained in a hyperplane through Θ_n, i.e., there must exist some $y \in \mathbb{R}^n$, $y \neq \Theta_n$, with

$$y^T x = 0 \qquad \text{for all} \quad x \in R(N).$$

This implies

$$\sum_{k=1}^{N} y^T A_{N-1} A_{N-2} \cdots A_k B_{k-1} u^k = 0 \qquad \text{for all} \quad \left(u^{1\,T}, \ldots, u^{N\,T}\right)^T \in (\mathbb{R}^m)^N,$$

hence

$$y^T A_{N-1} A_{N-2} \cdots A_k B_{k-1} = \Theta_m^T \qquad \text{for all} \quad k = 1, \ldots, N.$$

Since (2.98) also holds true for all $N \geq N_0$ instead of N_0, it follows that $y = \Theta_n$ which contradicts $y \neq \Theta_n$. Hence the assumption is false and the proof is complete. □

As a consequence of Theorem 2.25 we obtain

Theorem 2.26. *In addition to the assumption of Theorem 2.25 let*

$$\sup_{k \in \mathbb{N}_0} ||A_k|| < 1 \qquad \textit{where } ||\cdot|| \textit{ denotes the spectral norm.} \tag{2.100}$$

Then there is some $N \in \mathbb{N}$ and some $u \in U$ such that (2.98) holds true.

Proof. Assumption (2.100) implies

$$\lim_{N \to \infty} A_{N-1} A_{N-2} \cdots A_0 \, (x_0 - \widehat{x}) = \Theta_n.$$

Hence Theorem 2.25 implies that there is some $N \in \mathbb{N}$ with $N \geq N_0$ such that

$$-A_{N-1} A_{N-2} \cdots A_0 \, (x_0 - \widehat{x}) \in R(N)$$

which completes the proof. □

2.3.2 The General Problem of Controllability

We consider the same situation as at the beginning of Section 2.3.1. However, we do not assume the existence of a fixed point $\widehat{x} \in \mathbb{R}^n$ of the uncontrolled system (2.91), i.e., a solution of (2.89). Instead we assume a vector $x_1 \in \mathbb{R}^n$ to be given and consider the general

Problem of Controllability

Find some $N \in \mathbb{N}_0$ and a control function $u \in U$ (2.90) such that the solution $x : \mathbb{N}_0 \to \mathbb{R}^n$ of (2.86), (2.87) satisfies the end condition

$$x(N) = x_1. \tag{2.101}$$

From (2.86) and (2.87) we infer

$$\begin{aligned} x(N) &= g_{N-1}\left(g_{N-2}\left(\cdots\left(g_0\left(x_0, u(0)\right), u(1)\right), \ldots\right), u(N-1)\right) \\ &= G^N\left(x_0, u(0), \ldots, u(N-1)\right). \end{aligned} \tag{2.102}$$

Hence the end condition (2.101) is equivalent to

$$G^N\left(x_0, u(0), \ldots, u(N-1)\right) = x_1. \tag{2.103}$$

So we have to find vectors $u(0), \ldots, u(N-1) \in \Omega$ such that (2.103) is satisfied.

For every $N \in \mathbb{N}$ we define the controllable set

$$\begin{aligned} S_N\left(x_1\right) = \{x \in \mathbb{R}^n | \text{ there exists some } u \in U \\ \text{such that } G^N(x, u(0), \ldots, u(N-1)) = x_1\} \end{aligned}$$

and put

$$S\left(x_1\right) = \bigcup_{\mathbb{N} \in \mathbb{N}} S_N\left(x_1\right).$$

Now let $x_0 \in S\left(x_1\right)$. Then we ask the question under which conditions is x_0 an interior point of $S\left(x_1\right)$?

In order to find an answer to this question we assume that

$$\Omega \text{ is open} \quad \text{and} \quad g_N \in C^1\left(\mathbb{R}^n \times \mathbb{R}^m\right) \qquad \text{for all} \quad N \in \mathbb{N}_0.$$

Then it follows that $G^N \in C^1\left(\mathbb{R}^n \times \mathbb{R}^{m \cdot N}\right)$ for every $N \in \mathbb{N}$ and

$$\begin{aligned} & G_x^N\left(x, u(0), \ldots, u(N-1)\right) \\ & \quad = \left(g_{N-1}\right)_x\left(G^{N-1}\left(x, u(0), \ldots, u(N-2)\right), u(N-1)\right) \\ & \qquad \times \left(g_{N-2}\right)_x\left(G^{N-2}\left(x, u(0), \ldots, u(N-3)\right), u(N-2)\right) \\ & \qquad \times \qquad\qquad \vdots \\ & \qquad \times \left(g_0\right)_x\left(x, u(0)\right), \end{aligned}$$

and

$$\begin{aligned} &G^N_{u(k)}(x, u(0), \ldots, u(N-1)) \\ &\quad = (g_{N-1})_x \left(G^{N-1}(x, u(0), \ldots, u(N-2)), u(N-1)\right) \\ &\qquad \times (g_{N-2})_x \left(G^{N-2}(x, u(0), \ldots, u(N-3)), u(N-2)\right) \\ &\qquad \times \qquad \vdots \\ &\qquad \times (g_{k+1})_x \left(G^{k+1}(x, u(0), \ldots, u(k)), u(k+1)\right) \\ &\qquad \times (g_k)_{u(k)} \left(G^k(x, u(0), \ldots, u(k-1)), u(k)\right) \end{aligned}$$

for $k = 0, \ldots, N-1$, $x \in \mathbb{R}^n$ and $u \in U$.

Let us assume that $x_0 \in S_{N_0}(x_1)$ for some $N_0 \in \mathbb{N}$, i.e.,

$$G^{N_0}(x_0, u_0(0), \ldots, u_0(N_0-1)) = x_1$$

for some $u_0 \in U$. Further let $(g_N)_x(x, u_0)$ be non-singular for all $N \in \mathbb{N}_0$, for all $x \in \mathbb{R}^n$ and all $u \in \Omega$.

Then $G^{N_0}_x(x_0, u_0(0), \ldots, u_0(N_0-1))$ is also non-singular and, by the implicit function theorem, there exists an open set $V \subseteq \Omega^{N_0}$ with $(u_0(0), \ldots, u_0(N_0-1)) \in V$ and a function $h : V \to \mathbb{R}^n$ with $h \in C^1(V)$ such that

$$h(u_0(0), \ldots, u_0(N_0-1)) = x_0$$

and

$$G^{N_0}\left(h(u_0(0), \ldots, u_0(N_0-1)), u(0), \ldots, u(N_0-1)\right) = x_1 \quad \text{for all} \quad (u(0), \ldots, u(N_0-1)) \in V$$

which means

$$h(u(0), \ldots, u(N_0-1)) \in S_{N_0}(x_1) \qquad \text{for all} \quad (u(0), \ldots, u(N_0-1)) \in V.$$

Moreover,

$$\begin{aligned} h_{u(k)}(u_0(0), \ldots, u_0(N_0-1)) = -G^{N_0}_x&(x_0, u_0(0), \ldots, u_0(N_0-1))^{-1} \\ &\times G^{N_0}_{u(k)}(x_0, u_0(0), \ldots, u_0(N_0-1)). \end{aligned}$$

Next we assume that

$$\begin{aligned} \operatorname{rank}\big(h_{u(0)}&(u_0(0), \ldots, u_0(N_0-1)) \mid \ldots \\ &\mid h_{u(N_0-1)}(u_0(0), \ldots, u_0(N_0-1))\big) = n. \end{aligned}$$

Then it follows with the aid of the inverse function theorem that there exists an $n-$dimensional relatively open set $\widetilde{V} \subseteq V$ with $(u_0(0), \ldots, u_0(N_0 - 1)) \in \widetilde{V}$ such that the restriction of h to $\widetilde{V}$ is a homeomorphism which implies that $h\left(\widetilde{V}\right) \subseteq S_{N_0}(x_1)$ is open.

Therefore $x_0 \in h\left(\widetilde{V}\right)$ is an interior point of $S(x_1)$.

Now we consider the special case where there exists some $\widehat{x} \in \mathbb{R}^n$ with

$$g_N(\widehat{x}, \Theta_m) = \widehat{x} \qquad \text{for all} \quad N \in \mathbb{N}$$

which implies

$$G^N\left(\widehat{x}, \Theta_m^N\right) = \widehat{x} \qquad \text{for all} \quad N \in \mathbb{N}.$$

Then, for every $N \in \mathbb{N}$, it follows that $\widehat{x} \in S_N(\widehat{x})$, hence $\widehat{x} \in S(\widehat{x})$.

Let us assume that

$$(g_N)_x(\widehat{x}, \Theta_m) \text{ is non-singular} \qquad \text{for all} \quad N \in \mathbb{N}_0.$$

Then

$$G_x^N\left(\widehat{x}, \Theta_m^N\right) = (g_{N-1})_x(\widehat{x}, \Theta_m) \cdot (g_{N-2})_x(\widehat{x}, \Theta_m) \cdots (g_0)_x(\widehat{x}, \Theta_m)$$

is also non-singular for all $N \in \mathbb{N}$.

By the implicit function theorem we therefore conclude, for every $N \in \mathbb{N}$, that there exists an open set $V_N \subseteq \Omega^N$ with $\Theta_m^N \in V_N$ and a function $h_N : V_N \to \mathbb{R}^n$ with $h_N \in \mathcal{C}^1(V_N)$ such that

$$h_N\left(\Theta_m^N\right) = \widehat{x} \quad \text{and} \quad G^N\left(h_N(u(0), \ldots, u(N-1)), u(0), \ldots, u(N-1)\right) = \widehat{x}$$
$$\text{for all} \quad (u(0), \ldots, u(N-1)) \in V_N$$

which means

$$h_N(u(0), \ldots, u(N-1)) \in S_N(\widehat{x}) \qquad \text{for all} \quad (u(0), \ldots, u(N-1)) \in V_N.$$

Moreover,

$$(h_N)_{u(k)}\left(\Theta_m^N\right) = -G_x^N\left(\widehat{x}, \Theta_m^N\right)^{-1} \cdot G_{u(k)}^N\left(\widehat{x}, \Theta_m^N\right).$$

Next we assume that, for some $N_0 \in \mathbb{N}$,

$$\operatorname{rank}\left(\left(h_{N_0}\right)_{u(0)}\left(\Theta_m^{N_0}\right) \mid \ldots \mid \left(h_{N_0}\right)_{u(N_0-1)}\left(\Theta_m^{N_0}\right)\right) = n.$$

Then it follows with the aid of the inverse function theorem that there exists an $n-$dimensional relatively open set $\widetilde{V}_{N_0} \subseteq V_{N_0}$ with $\Theta_m^{N_0} \in \widetilde{V}_{N_0}$ such that the restriction of h_{N_0} to $\widetilde{V}_{N_0}$ is a homeomorphism which implies that $h_{N_0}\left(\widetilde{V}_{N_0}\right) \subseteq S_{N_0}(\widehat{x})$ is open. Therefore $\widehat{x} \in h_{N_0}\left(\widetilde{V}_{N_0}\right)$ is an interior point of $S(\widehat{x})$. This result is a generalization of Theorem 2.25, if Ω in addition is open.

2.3.3 Stabilization of Controlled Systems

Let $(g_t)_{t\in\mathbb{N}}$ be a sequence of continuous mappings $g_t : \mathbb{R}^n \times \mathbb{R}^m \to \mathbb{R}^n$ and let $\mathcal{H}$ be a family of continuous mappings $h : \mathbb{R}^n \to \mathbb{R}^m$. If we define, for every $h \in \mathcal{H}$ and $t \in \mathbb{N}$, the mapping $f_t^h : \mathbb{R}^n \to \mathbb{R}^n$ by

$$f_t^h(x) = g_t(x, h(x)), \qquad x \in \mathbb{R},$$

then we obtain a non-autonomous time-discrete dynamical system $\left(\mathbb{R}^n, \left(f_t^h\right)_{t\in\mathbb{N}}\right)$. The dynamics in this system is defined by the sequence $F^h = \left(F_t^h\right)_{t\in\mathbb{N}}$ of mappings $F_t^h : \mathbb{R}^n \to \mathbb{R}^n$ given by

$$F_t^h(x) = f_t^h \circ f_{t-1}^h \circ \ldots \circ f_1^h(x) \qquad \text{for all} \quad x \in \mathbb{R}^n \quad \text{and} \quad t \in \mathbb{N}$$

and

$$F_0^h(x) = x \qquad \text{for all} \quad x \in \mathbb{R}^n.$$

We also obtain the dynamical system $\left(\mathbb{R}^n, \left(f_t^h\right)_{t\in\mathbb{N}}\right)$, if we replace the control function $u : \mathbb{N}_0 \to \mathbb{R}^m$ in the system (2.86) by the feedback controls $h(x) : \mathbb{N}_0 \to \mathbb{R}^m$, $x \in \mathbb{R}^n$.

The problem of stabilization of the controlled system (2.86) by the feedback controls $h(x)$, $x \in \mathbb{R}^n$, then reads as follows: Given $x_0 \in \mathbb{R}^n$ such that the limit set $L_{F^h}(x_0)$ defined by (1.70) (see Section 1.3.8) is non-empty and compact for all $h \in \mathcal{H}$.

Find a mapping $h \in \mathcal{H}$ such that $L_{F^h}(x_0)$ is stable, an attractor or asymptotically stable with respect to $\left(f_t^h\right)_{t\in\mathbb{N}}$.

Let us consider the special case

$$g_t(x,u) = A_t(x)\,x + B_t(x)\,u \qquad \text{for} \quad x \in \mathbb{R}^n,\; u \in \mathbb{R}^m, \tag{2.104}$$

where $(A_t(x))_{t\in\mathbb{N}}$ and $(B_t(x))_{t\in\mathbb{N}}$ is a sequence of real, continuous $n \times n-$ and $n \times m-$matrix functions on $\mathbb{R}^n$, respectively.

Let $\mathcal{H}$ be the family of all linear mappings $h : \mathbb{R}^n \to \mathbb{R}^m$ (which are automatically continuous). Every $h \in \mathcal{H}$ is then representable in the form

$$h(x) = C^h\,x, \qquad x \in \mathbb{R}^n,$$

where C^h is a real $m \times n-$matrix. For every $t \in \mathbb{N}$ and $h \in \mathcal{H}$ we therefore obtain

$$f_t^h(x) = \left(A_t(x) + B_t(x)\,C^h\right) x, \qquad x \in \mathbb{R}^n. \tag{2.105}$$

Let us put

$$D_t^h(x) = A_t(x) + B_t(x)\,C^h \qquad \text{for all} \quad x \in \mathbb{R}^n.$$

If we choose $x_0 \in \Theta_n$ = zero vector of $\mathbb{R}^n$, then we conclude

$$F_t^h(x_0) = x_0 \qquad \text{for all} \quad t \in \mathbb{N}_0,\ h \in \mathcal{H},$$

and therefore $L_{F^h}(x_0) = \{x_0\}$.

The problem of stabilization of the controlled system (2.86) with g_t, $t \in \mathbb{N}$, given by (2.104) in this situation consists of finding an $m \times n-$matrix C^h such that $\{x_0 = \Theta_n\}$ is stable, an attractor or asymptotically stable with respect to $(f_t^h)_{t\in\mathbb{N}}$ with f_t^h given by (2.105).

Now let us assume that

$$||D_t^h(x)|| \leq 1 \qquad \text{for all} \quad x \in \mathbb{R}^n \quad \text{and} \quad t \in \mathbb{N} \tag{2.106}$$

where $||\cdot||$ denotes the spectral norm.

Let $U \subseteq \mathbb{R}^n$ be a relatively compact open set with $x_0 = \Theta_n \in U$. Then there is some $r > 0$ such that

$$B_U = \{x \in \mathbb{R}^n \mid ||x||_2 < r\} \subseteq U.$$

Hence B_U is open, $x_0 \in B_U$, and assumption (2.106) implies

$$f_t^h(B_U) \subseteq B_U \qquad \text{for all} \quad t \in \mathbb{N}.$$

If we define

$$V(x) = ||x||_2^2 = x^T x \quad \text{for} \quad x \in \mathbb{R}^n,$$

then

$$V(x) \geq 0 \qquad \text{for all} \quad x \in \mathbb{R}^n \qquad \text{and} \qquad (V(x) = 0 \Leftrightarrow x = x_0 = \Theta_n)$$

and

$$\begin{aligned} V\left(f_t^h(x)\right) - V(x) &= x^T D_t^h(x)^T D_t^h(x)\, x - x^T x \\ &= ||D_t^h(x)\, x||_2^2 - ||x||_2^2 \leq \left(||D_t^h(x)\, x|| - 1\right) ||x||_2^2 \leq 0 \end{aligned}$$

for all $x \in \mathbb{R}^n$ and $t \in \mathbb{N}$.

This shows that V is a Lyapunov function with respect to $(f_t^h)_{t\in\mathbb{N}}$ on $G = \mathbb{R}^n$ which is positive definite with respect to $\{x_0 = \Theta_n\}$. By Theorem 1.9 we therefore conclude that $\{x_0 = \Theta_n\}$ is stable with respect to $(f_t^h)_{t\in\mathbb{N}}$ with f_t^h given by (2.105).

Next we assume that

$$\sup_{t\in\mathbb{N}} ||D_t^h(x)|| < 1 \qquad \text{for all} \quad x \in \mathbb{R}^n. \tag{2.107}$$

Then it follows from

$$\begin{aligned} V\left(F_t^h(x)\right) &= x^T D_1^h(x)^T \cdots D_t^h\left(F_{t-1}^h(x)\right)^T D_t^h\left(F_{t-1}^h(x)\right) \cdots D_1^h(x)\, x \\ &= \|D_t^h\left(F_{t-1}^h(x)\right) \cdots D_1^h(x)\, x\|_2^2 \\ &\le \|D_t^h\left(F_{t-1}^h(x)\right) x\|^2 \cdots \|D_1^h(x)\, x\|^2 \|x\|^2 \end{aligned}$$

for all $x \in \mathbb{R}^n$ and $t \in \mathbb{N}$ that

$$\lim_{t\to\infty} V\left(F_t^h(x)\right) = 0 \qquad \text{for all} \quad x \in \mathbb{R}^n.$$

This implies

$$\lim_{t\to\infty} F_t^h(x) = \Theta_n \qquad \text{for all} \quad x \in \mathbb{R}^n$$

and shows that $\{x_0 = \Theta_n\}$ is an attractor with respect to $\left(f_t^h\right)_{t\in\mathbb{N}}$ with f_t^h given by (2.105).

Result. *Under the assumption (2.107) the set $\{x_0 = \Theta_n\}$ is asymptotically stable with respect to $\left(f_t^h\right)_{t\in\mathbb{N}}$ with f_t^h given by (2.105).*

2.3.4 The Problem of Reachability

We again consider the situation at the beginning of Section 2.3.1 without necessarily assuming the existence of a fixed point $\widehat{x} \in \mathbb{R}^n$ of the uncontrolled system (2.91). Let $\Omega \subseteq \mathbb{R}^m$ be a non-empty subset. For a given $x_0 \in \mathbb{R}^n$ we then define the set of states that are reachable from x_0 in $N \in \mathbb{N}$ steps by

$$R_N(x_0) = \Big\{x = G^N\left(x_0, u(0), \ldots, u(N-1)\right) \;\Big|\; u(k) \in \Omega,\ k = 0, \ldots, N-1\Big\} \tag{2.108}$$

where the map $G^N : \mathbb{R}^{m\cdot N} \to \mathbb{R}^n$ is defined by (2.102).

Further we define the set of states reachable from x_0 in a suitable number of steps by

$$R(x_0) = \bigcup_{N\in\mathbb{N}} R_N(x_0). \tag{2.109}$$

The question we are interested in now is: Under which conditions does $R(x_0)$ have a non-empty interior?

A simple answer to this question gives

Theorem 2.27. *Let Ω be open. If there is some $N \in \mathbb{N}$ and there exist $u(0), \ldots, u(N-1) \in \Omega$ such that*

$$\operatorname{rank}\Big(G^N_{u(0)}\,(x_0, u(0), \ldots, u(N-1)) \mid \ldots \mid G^N_{u(N-1)}\,(x_0, u(0), \ldots, u(N-1))\Big) = n, \qquad (2.110)$$

then $R_N\,(x_0)$ has a non-empty interior and therefore also $R\,(x_0)$.

Proof. Condition (2.110) implies that the $n \times N \cdot m-$matrix

$$\Big(G^N_{u(0)}\,(x_0, u(0), \ldots, u(N-1)) \mid \ldots \mid G^N_{u(N-1)}\,(x_0, u(0), \ldots, u(N-1))\Big)$$

has n linearly independent column vectors. Let E be the $n-$dimensional subset of Ω^N consisting of all vectors whose components which do not correspond to these linearly independent column vectors are equal to the ones of $\left(u(0)^T, \ldots, u(N-1)^T\right)^T$. If we restrict the mapping G^N to E, then the Jacobi matrix of this restriction consists of these linearly independent column vectors and is therefore non-singular.

By the inverse function theorem therefore there exists an open set (with respect to E) $U \subseteq \Omega^N$ with $\left(u(0)^T, \ldots, u(N-1)^T\right)^T \in U$ which is mapped homeomorphically by G^N on an open set $V \subseteq R_N\,(x_0)$ with $G^N\,(x_0, u(0), \ldots, u(N-1)) \in V$. This completes the proof. □

Next we consider the linear case where

$$g_t(x, u) = A_t\, x + B_t\, u, \qquad x \in \mathbb{R}^n,\ u \in \mathbb{R}^m,$$

with $n \times n-$matrices A_t and B_t, respectively, for every $t \in \mathbb{N}_0$.

Then, for every $N \in \mathbb{N}$ and every $x_0 \in \mathbb{R}^n$, we obtain

$$G^N\,(x_0, u(0), \ldots, u(N-1)) = A_{N-1} \cdots A_0\, x_0 + \sum_{k=1}^{N} A_{N-1} \cdots A_k B_{k-1}\, u(k-1)$$

where for $k = N$ we put $A_{N-1} \cdots A_k = I = n \times n-$unit matrix. Further we have, for every $N \in \mathbb{N}$ and every $x_0 \in \mathbb{R}^n$,

$$R_N\,(x_0) = \Big\{x = A_{N-1} \cdots A_0\, x_0 + \sum_{k=1}^{N} A_{N-1} \cdots A_k B_{k-1}\, u(k-1) \;\Big|\; u(k) \in \Omega,\ k = 0, \ldots, N-1\Big\}.$$

Because of

$$G^N_{u(k-1)}\left(x_0, u(0), \ldots, u(N-1)\right) = A_{N-1} \cdots A_k B_{k-1} \qquad \text{for} \quad k = 1, \ldots, N$$

it follows that the condition (2.110) for $N = N_0$ coincides with the condition (2.98).

If this is satisfied, then by Theorem 2.27 the set $R\left(x_0\right)$ (2.109) of states reachable from x_0 has a non-empty interior, if Ω is open.

If $\Omega = \mathbb{R}^m$, it follows in addition that $R\left(x_0\right) = \mathbb{R}^n$ for all $x_0 \in \mathbb{R}^n$.

Proof. Let $x, x_0 \in \mathbb{R}^n$ be given arbitrarily. Then condition (2.98) implies the existence of $u(k) \in \mathbb{R}^m$ for $k = 0, \ldots, N_0 - 1$ such that

$$x - A_{N_0-1} \cdots A_0\, x_0 = \sum_{k=1}^{N_0} A_{N_0-1} \cdots A_k B_{k-1}\, u(k-1)$$

holds true which shows that $x \in R_{N_0}\left(x_0\right) \subseteq R\left(x_0\right)$. □

For every $k = 1, \ldots, N$ let us define an $n \times m-$matrix C^k by

$$C^k = A_{N-1} \cdots A_k B_{k-1} \qquad \text{for} \quad k = 1, \ldots, N-1$$

and

$$C^N = B_{N-1}.$$

The condition (2.98) implies the existence of n column vectors

$$\begin{pmatrix} c^{k_l}_{1 j_{k_l}} \\ \vdots \\ c^{k_l}_{n j_{k_l}} \end{pmatrix} \qquad \text{for} \quad l = 1, \ldots, n \quad \text{which are linearly independent.}$$

Let us define the $n \times n-$matrix C and a vector $u \in \mathbb{R}^n$ by

$$C = \begin{pmatrix} c^{k_1}_{1 j_{k_1}} & \cdots & c^{k_n}_{1 j_{k_n}} \\ & \vdots & \\ c^{k_1}_{n j_{k_1}} & \cdots & c^{k_n}_{n j_{k_n}} \end{pmatrix} \qquad \text{and} \qquad u = \begin{pmatrix} u_{j k_l}\left(k_1 - 1\right) \\ \vdots \\ u_{j k_n}\left(k_n - 1\right) \end{pmatrix}, \text{ respectively,}$$

and put

$$u_j(k-1) = 0 \qquad \text{for} \quad k \neq k_l,\ j \neq j_{k_l},\ l = 1, \ldots, n.$$

Then we obtain

$$G^N(x_0, u(0), \ldots, u(N-1)) = A_{N-1} \cdots A_0 x_0 + C u$$

which implies

$$u = C^{-1}\Big(\underbrace{G^N(x_0, u(0), \ldots, u(N-1))}_{=x} - A_{N-1} \cdots A_0 x_0\Big).$$

Now let

$$E = \Big\{u = (u(0), \ldots, u(N-1)) \in \mathbb{R}^{m \cdot N} \\ \Big| \; u_j(k-1) = 0 \text{ for } k \neq k_l \text{ and } j \neq j_{k_l}, \; l = 1, \ldots, n\Big\}.$$

Then $G^N(x_0, \cdot)$ is a linear isomorphism from E on $\mathbb{R}^n$.

Therefore

$$G^N(x_0, u) = x \qquad \text{for some} \quad u \in E$$

implies

$$u = G^N(x_0, \cdot)^{-1}(x),$$
$$G^N(x_0, u) = A_{N-1} \cdots A_0 x_0 + C \cdot G^N(x_0, \cdot)^{-1}(x).$$

If all $A_k, \; k \in \mathbb{N}_0$ are invertible it follows that

$$x_0 = A_0^{-1} \cdot A_{N-1}^{-1} x - A_0^{-1} \cdot A_{N-1}^{-1} C \cdot G^N(x_0, \cdot)^{-1}(x).$$

In the nonlinear case we have the following situation:

If the condition (2.110) is satisfied, there exists an $n-$dimensional subset E of Ω^N and a set $U \subseteq E$ which is open with respect to E and contains $\left(u(0)^T, \ldots, u(N-1)^T\right)^T$ and which is mapped homeomorphically on an open $V \subseteq R_N(x_0)$ by the restriction of $G^N(x_0, \cdot)$ to E. If

$$x = G^N(x_0, u(0), \ldots, u(N-1)),$$

then

$$\left(u(0)^T, \ldots, u(N-1)^T\right)^T = G^N(x_0, \cdot)^{-1}(x).$$

If in addition $G_x^N(x_0, u(0), \ldots, u(N-1))$ is non-singular, then by the implicit function theorem there exists an open set $W \subseteq \Omega^N$ which contains $\left(u(0)^T, \ldots, u(N-1)^T\right)^T$ and a function $h : W \to \mathbb{R}^n$ with $h \in \mathcal{C}^1(W)$ such that

$$h(u(0), \ldots, u(N-1)) = x_0$$

and

$$G^N\left(h\left(\widetilde{u}(0),\ldots,\widetilde{u}(N-1)\right),\widetilde{u}(0),\ldots,\widetilde{u}(N-1)\right)=x$$
$$\text{for all}\quad \left(\widetilde{u}(0)^T,\ldots,\widetilde{u}(N-1)^T\right)^T\in W.$$

This implies

$$x_0=h\left(G^N\left(x_0,\cdot\right)^{-1}(x)\right).$$

Since

$$\begin{aligned}h_{u(k)}\left(u(0),\ldots,u(N-1)\right)=-G_x\left(x_0,u(0),\ldots,u(N-1)\right)^{-1}\\ \times G^{N_0}_{u(k)}\left(x_0,u(0),\ldots,u(N-1)\right)\\ \text{for}\quad k=0,\ldots,N-1,\end{aligned}$$

it follows from (2.110) that

$$\operatorname{rank}\left(h_{u(0)}\left(u(0),\ldots,u(N-1)\right)\mid\ldots\mid h_{u(N-1)}\left(u(0),\ldots,u(N-1)\right)\right)=n$$

which implies that h maps $U\cap W$ homeomorphically onto an open set $\widetilde{V}\subseteq R_N(x_0)$ which contains x. Therefore $h\circ G^N(x_0,\cdot)^{-1}$ maps $V\cap\widetilde{V}$ homeomorphically on an open set $\widetilde{\widetilde{V}}$ which contains x_0 and is contained in

$$\begin{aligned}S_N(x)=\{\widetilde{x}\in\mathbb{R}^n\mid\text{there exists some }\widetilde{u}\in U\ (2.53)\\ \text{with }G^N\left(\widetilde{x},\widetilde{u}(0),\ldots,\widetilde{u}(N-1)\right)=x\}.\end{aligned}$$

Finally let us assume (as in Theorem 2.27) that Ω is open and (for given $x_0\in\mathbb{R}^n$) there exists some $N\in\mathbb{N}$ such that the condition (2.110) is satisfied for all $\left(u(0)^T,\ldots,u(N-1)^T\right)^T\in\Omega^N$. Then it follows from the proof of Theorem 2.27 that every $x\in R_N(x_0)$ is an interior point of $R_N(x_0)$, i.e., $R_N(x_0)$ is open.

This implies in the linear case with Ω being an open subset of $\mathbb{R}^m$ that $R_N(x_0)$ is open for every $x_0\in\mathbb{R}^n$, if the condition (2.98) is satisfied.

3

Chaotic Behavior of Autonomous Time-Discrete Systems

3.1 Chaos in the Sense of Devaney

Let $f : X \to X$ be a continuous mapping of a metric space X into itself. Then by the definition

$$\pi(x, n) = f^n(x) \qquad \text{for all} \quad x \in X \quad \text{and} \quad n \in \mathbb{N}_0 \tag{3.1}$$

with

$$f^0(x) = x \qquad \text{and} \qquad f^{n+1}(x) = f\left(f^n(x)\right), \quad n \in \mathbb{N}_0, \tag{3.2}$$

we obtain an autonomous time-discrete dynamical system (see Section 1.3.1).

A point $x \in X$ is called period point of $f : X \to X$, if there exists an $n \in \mathbb{N}$ with $f^n(x) = x$. The smallest $n \in \mathbb{N}$ with this property is called $f-$period of $x \in X$.

Definition. *A continuous mapping $f : X \to X$ is called topologically transitive, if for every pair of non-empty open subsets $U, V \subseteq X$ there is an $n \in \mathbb{N}$ with $f^n(U) \cap V \neq \emptyset$ where*

$$f^n(U) = \{f^n(x) \mid x \in U\}.$$

Graphically this means that a topologically transitive mapping $f : X \to X$ causes a complete mixing of the metric space X, if it is applied sufficiently often.

On using the two concepts "period points" and "topological transitivity" we now define chaos as follows:

Definition. *A continuous mapping f of a metric space X into itself is called chaotic, if the following two conditions are satisfied:*
(1) The set of period points of f is dense in X.
(2) The mapping $f : X \to X$ is topologically transitive.

S. Pickl and W. Krabs, *Dynamical Systems: Stability, Controllability and Chaotic Behavior*, 149
DOI 10.1007/978-3-642-13722-8_3, © Springer-Verlag Berlin Heidelberg 2010

Every period point $x \in X$ of f can be assigned an orbit

$$O(x) = \{f^k(x) \mid k = 0, \ldots, n\} \qquad \text{for some} \quad n \in \mathbb{N}_0$$

whose points are also period points of f.

Each such "periodic" orbit can be considered as an ordered movement in the time-discrete dynamical system that is defined by f. Every period point therefore defines a certain order; the topological transitivity, however, is the expression of complete disorder. The above chaos definition therefore states that order and complete disorder are immediately neighboured, if the mapping $f : X \to X$ is chaotic.

This definition goes back to Devaney (see [5]) who adds a third property which is defined as follows:

Definition. *A continuous mapping $f : X \to X$ is called sensitively depending on initial values, if there exists some $\delta > 0$ such that for every $x \in X$ and every neighborhood $V(x)$ of x there exists some $y \in V(x)$ and some $n \in \mathbb{N}_0$ with*

$$d\left(f^n(x), f^n(y)\right) \geq \delta$$

where d denotes the metric of X.

This property, however, is superfluous, if X consists of infinitely many elements, because of the

Theorem 3.1. *Let the metric space X consist of infinitely many elements and let $f : X \to X$ be chaotic in the sense of the above definition. Then f is also sensitively depending on initial values (see [2]).*

Proof. Since X consists of infinitely many points, there are two period points $x_1, x_2 \in X$ of f with $O(x_1) \neq O(x_2)$ where

$$O(x) = \{f^n(x) \mid n \in \mathbb{N}_0\} \qquad \text{for every} \quad x \in X.$$

Let

$$\delta_0 = d\left(O(x_1), O(x_2)\right) = \min\left\{d(y_1, y_2) \mid y_i \in O(x_i),\ i = 1, 2\right\}.$$

Let $x \in X$ be chosen arbitrarily. Then either

(a) $d\left(x, O(x_1)\right) = \min\left\{d(x, y) \mid y \in O(x_1)\right\} \geq \dfrac{\delta_0}{2}$ or

(b) $d\left(x, O(x_1)\right) < \dfrac{\delta_0}{2}$.

Let $d(x, \widehat{y}_1) = d\left(x, O(x_1)\right)$. Then it follows in case (b), for every $y_2 \in O(x_2)$, the chain of inequalities

$$d(x, y_2) \geq d(\widehat{y}_1, y_2) - d(x, \widehat{y}_1) > \delta_0 - \frac{\delta_0}{2} = \frac{\delta_0}{2}, \text{ hence } d\left(x, O(x_2)\right) > \frac{\delta_0}{2}.$$

Result. *There is a number* $\delta_0 > 0$ *such that for every* $x \in X$ *there exists a period point* $\widehat{x} \in X$ *with* $d\left(x, O\left(\widehat{x}\right)\right) \geq \dfrac{\delta_0}{2}$.

Let us put $\delta = \dfrac{\delta_0}{8}$. Then we choose $x \in X$ arbitrarily and an arbitrary neighbourhood $V(x)$. Further we put

$$U = V(x) \cap B_\delta(x) \qquad \text{where} \quad B_\delta(x) = \{y \in X \mid d(y,x) < \delta\}.$$

By property (1) there is a period point $p \in U$ of f. Let $f^n(p) = p$. By the above result there is a period point $q \in X$ of f with $d(x, O(q)) \geq \dfrac{\delta_0}{2} = 4\delta$.

We put

$$V = \bigcap_{i=0}^{n} f^{-i}\left(B_\delta\left(f(q)\right)\right)$$

where, for every non-empty subset W of X, we have

$$f^{-i}(W) = \left\{y \in X \mid f^i(y) \in W\right\}, \qquad i \in \mathbb{N}_0.$$

Then V is open and non-empty, since $q \in V$.

Since f is topologically transitive, there is some $y \in U$ and some $k \in \mathbb{N}$ with $f^k(y) \in V$.

Now let $j = \left[\dfrac{k}{n} + 1\right] \;\Rightarrow\; 1 \leq n \cdot j - k \leq n$. By construction we therefore have

$$f^{n \cdot j}(y) = f^{n \cdot j - k}\left(f^k(y)\right) \in f^{n \cdot j - k}(V) \subseteq B_\delta\left(f^{n \cdot j - k}(q)\right).$$

Now we have $f^{n \cdot j}(p) = p$ and hence

$$\begin{aligned} d\left(f^{n \cdot j}(p), f^{n \cdot j}(y)\right) &= d\left(p, f^{n \cdot j}(y)\right) \\ &\geq d\Big(x, \underbrace{f^{n \cdot j - k}(q)}_{\in\, O(q)}\Big) - d\left(f^{n \cdot j - k}(q), f^{n \cdot j}(y)\right) - d(p, x) \\ &\geq 4\delta - \delta - \delta = 2\delta \end{aligned}$$

because of $p \in B_\delta(x)$. Therefore we further have either
(α) $d\left(f^{n \cdot j}(x), f^{n \cdot j}(y)\right) \geq \delta$ or
(β) $d\left(f^{n \cdot j}(x), f^{n \cdot j}(y)\right) < \delta$
where in the latter case we conclude that

$$d\left(f^{n \cdot j}(x), f^{n \cdot j}(p)\right) \geq d\left(f^{n \cdot j}(p), f^{n \cdot j}(y)\right) - d\left(f^{n \cdot j}(y), f^{n \cdot j}(x)\right) > 2\delta - \delta = \delta.$$

Because of $y \in U$ and $p \in U$ we have found in each case a $z \in V(x)$ with $d\left(f^{\,n \cdot j}(x), f^{\,n \cdot j}(z)\right) \geq \delta$ which completes the proof. □

A paradigmatic example for a chaotic mapping is the so called shift-mapping in the space of $0-1$–sequences, i.e., in the space

$$\Sigma = \{s = (s_0, s_1, s_2, \ldots) \mid s_j \in \{0,1\} \quad \text{for} \quad j \in \mathbb{N}_0\}.$$

If one defines a function $d : \Sigma \times \Sigma \to \mathbb{R}_+$ by

$$d(s,t) = \sum_{i=0}^{\infty} \frac{|s_i - t_i|}{2^i}, \qquad s,t \in \Sigma,$$

then d turns out to be a metric.

In the following we need the

Lemma 3.2. *Let $s,t \in \Sigma$ be given. Then it follows that*
(a) $s_i = t_i$ *for* $i = 0, \ldots, n \implies d(s,t) \leq 2^{-n}$.
Conversely it follows that
(b) $d(s,t) < 2^{-n} \implies s_i = t_i$ *for* $i = 0, \ldots, n$.

Proof.
(a) Let $s_i = t_i$ for $i = 0, \ldots, n$. Then

$$d(s,t) = \sum_{i=n+1}^{\infty} \frac{|s_i - t_i|}{2^i} \leq \sum_{i=n+1}^{\infty} \frac{1}{2^i} = \frac{1}{2^{n+1}} \sum_{i=0}^{\infty} \frac{1}{2^i} = \frac{1}{2^n}.$$

(b) Let $d(s,t) < \frac{1}{2^n}$ for some $n \in \mathbb{N}_0$. If $s_i \neq t_i$ for some $i \in \{0, \ldots, n\}$, then it follows that $d(s,t) \geq 2^{-i} \geq 2^{-n}$.

So the implication (b) follows by contraposition. □

The shift-mapping $\sigma : \Sigma \to \Sigma$ is now defined by

$$\sigma(s_0, s_1, s_2, \ldots) = (s_1, s_2, s_3, \ldots). \tag{3.3}$$

At first we prove

Theorem 3.3. *The shift-mapping $\sigma : \Sigma \to \Sigma$ defined by (3.3) is continuous.*

Proof. Let $s = (s_0, s_1, s_2, \ldots) \in \Sigma$ and $\varepsilon > 0$ be given. Then we choose $n \in \mathbb{N}_0$ such that $\frac{1}{2^n} < \varepsilon$. Let $\delta = \frac{1}{2^{n+1}}$. If we then, for some $t = (t_0, t_1, t_2, \ldots) \in \Sigma$, have that $d(s,t) < \delta$, it follows from Lemma 3.2 that

$$s_i = t_i \qquad \text{for} \quad i = 0, \ldots, n+1,$$

hence $\sigma(s)_i = \sigma(t)_i$ for $i = 1, \ldots, n$ which again by Lemma 3.2 implies that $d(\sigma(s), \sigma(t)) \leq \frac{1}{2^n} < \varepsilon$. This completes the proof. □

For every $n \in \mathbb{N}_0$ there are 2^n different points $s \in \Sigma$ of the form

$$s = (s_0, \ldots, s_{n-1}, s_0, \ldots, s_{n-1}, \ldots)$$

for which it follows that $\sigma^n(s) = s$ which shows that they are period points of σ.

Next we prove

Theorem 3.4. *The set Per(σ) of all period points of σ is dense in Σ.*

Proof. Let $s = (s_0, s_1, s_2, \ldots) \in \Sigma$ be given arbitrarily. Then we define for every $n \in \mathbb{N}_0$ the sequence

$$\tau_n = (s_0, s_1, \ldots, s_n, s_0, s_1, \ldots, s_n, \ldots), \qquad n \in \mathbb{N}_0.$$

Therefore we have

$$\tau_n \in \text{Per}(\sigma) \qquad \text{for all} \quad n \in \mathbb{N}_0$$

and

$$\tau_{ni} = s_i \qquad \text{for} \quad i = 0, \ldots, n.$$

The Lemma 3.2 implies $d(\tau_n, s) \leq \frac{1}{2^n}$.

Let $\varepsilon > 0$ be given. Then we choose $n \in \mathbb{N}_0$ such that $\frac{1}{2^n} \leq \varepsilon$ and conclude that $d(\tau_n, s) \leq \varepsilon$ which completes the proof. □

Theorem 3.5. *There is some $\widehat{s} \in \Sigma$ such that the orbit $(\sigma^n(\widehat{s}))_{n \in \mathbb{N}_0}$ is dense in Σ.*

Proof. Let us define

$$\widehat{s} = (01\ \underbrace{00011011}_{\text{double blocs}}\ \underbrace{000001\ldots111}_{\text{triple blocs}}\ldots).$$

Obviously $\sigma^n(\widehat{s})$ coincides, for sufficiently large n, with every $t \in \Sigma$ in the first n bits. This implies by the Lemma 3.2 that $d(\sigma^n(\widehat{s}), t) \leq \frac{1}{2^n}$ and shows the assertion. □

As a consequence we obtain the

Theorem 3.6. *The shift-mapping (3.3) is topologically transitive and therefore chaotic.*

Proof. Let $U, V \subseteq \Sigma$ be open subsets of Σ. Take any $\widehat{s} \in U$. Then there is some $\delta_1 > 0$ with $B_s(\delta_1) = \{t \in \Sigma \mid d(t, \widehat{s}) < \delta_1\} \subseteq U$. Theorem 3.5 implies the existence of some $n \in \mathbb{N}_0$ such that $\sigma^n(\widehat{s}) \in B_s(\delta_1)$. Now take any $t \in V$.

Then there is some $\delta_2 > 0$ with $B_t(\delta_2) \subseteq V$ and there exists some $m \in \mathbb{N}_0$ such that $\sigma^{n+m}(\widehat{s}) \in B_t(\delta_2)$. Since $\sigma^{n+m}(\widehat{s}) \in \sigma^m(B_s(\delta_1))$, it follows that

$$\sigma^m(B_s(\delta_1)) \cap B_t(\delta_2) \neq \emptyset \quad \Longrightarrow \quad \sigma^m(U) \cap V \neq \emptyset$$

which completes the proof. □

3.2 Topological Conjugacy

Let X and Y be two metric spaces and let $f : X \to X$ and $g : Y \to Y$ be two continuous mappings.

Definition. *f and g are called topologically conjugated, if there exists a homeomorphism $h : X \to Y$ such that*

$$h \circ f = g \circ h, \tag{3.4}$$

i.e., if the following diagram is commutative

$$\begin{array}{ccc} X & \xrightarrow{f} & X \\ {\scriptstyle h}\downarrow & & \downarrow{\scriptstyle h} \\ Y & \xrightarrow[g]{} & Y \end{array}$$

Theorem 3.7. *Let $f : X \to X$ and $g : Y \to Y$ be continuous mappings of metric spaces X and Y, respectively, and let f be chaotic. If then there exists a continuous mapping $h : X \to Y$ which is surjective and for which (3.4) holds true, g is also chaotic.*

Proof. Let $y \in Y$ be an arbitrary point and let $W(y)$ be an arbitrary neighbourhood of y. Then there is some $x \in X$ with $h(x) = y$ and $h^{-1}(W(y)) = V(x)$ is a neighborhood of x. Since f is chaotic, there is a period point $p \in V(x)$ of f with $f^n(p) = p$ for some $n \in \mathbb{N}$. If we put $q = h(p)$, then $q \in W(y)$ and

$$g^n(q) = g^n(h(p)) = g^{n-1}(g(h(p))) = g^{n-1}(h(f(p))) = h\left(f^n(p)\right) = h(p) = q,$$

i.e., q is a period point of g. The set of period points is therefore also dense in Y.

Now let V and W be two non-empty open subsets of Y. Then $h^{-1}(V)$ and $h^{-1}(W)$ are two non-empty open subsets of X. Since f is chaotic, there exists some $n \in \mathbb{N}$ with $f^n\left(h^{-1}(V)\right) \cap h^{-1}(W) \neq \emptyset$. Now we have

$$\begin{aligned} g^n(V) = g^n \circ h\left(h^{-1}(V)\right) &= g^{n-1} \circ g \circ h\left(h^{-1}(V)\right) = g^{n-1} \circ h \circ f\left(h^{-1}(V)\right) \\ = \ldots \quad &= \quad h \circ f^n\left(h^{-1}(V)\right) \end{aligned}$$

which implies $g^n(V) \cap W \neq \emptyset$. Hence g is also topologically transitive and therefore chaotic which completes the proof. □

Corollary 3.8. *Let the continuous mappings $f : X \to X$ and $g : Y \to Y$ be topologically conjugated. Then f is chaotic, if and only if g is chaotic.*

Proof. If f is chaotic, then by Theorem 3.7 also g is chaotic. If conversely g is chaotic, then, since $h^{-1} : Y \to X$ is continuous and surjective and $h^{-1} \circ f = g \circ h^{-1}$, again by Theorem 3.7 f is also chaotic. □

Let us demonstrate this Corollary by an example. For that purpose we consider at first the mapping $f : \mathbb{R} \to \mathbb{R}$ which is defined by

$$f(x) = \mu\, x(1-x), \qquad x \in \mathbb{R}. \tag{3.5}$$

We assume that $\mu > 4$. Then the maximum value $f\left(\frac{1}{2}\right) = \frac{\mu}{4}$ of f is larger than 1 and there is an open interval $A_0 \subseteq [0,1]$ with $\frac{1}{2}$ as midpoint such that

$$f(x) > 1, \quad f^2(x) < 0 \qquad \text{and} \qquad \lim_{n\to\infty} f^n(x) = -\infty$$

for all $x \in A_0$. If one defines

$$A_1 = \{x \in [0,1] \mid f(x) \in A_0\},$$

then it follows that

$$f^2(x) > 1, \quad f^3(x) < 0 \qquad \text{and} \qquad \lim_{n\to\infty} f^n(x) = -\infty$$

for all $x \in A_1$. Graphically we have the following situation:

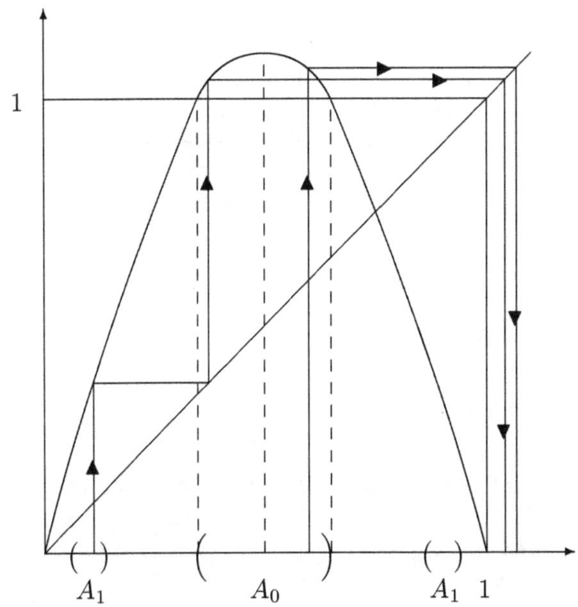

Next we define

$$A_2 = \{x \in [0,1] \mid f^i(x) \in A_{i-1} \text{ for } i = 1,2\}.$$

Then it follows that $f^3(x) > 1$ for all $x \in A_2$.

If one defines, for an arbitrary $n \in \mathbb{N}$,

$$A_n = \{x \in [0,1] \mid f^i(x) \in A_{i-1} \text{ for } i = 1,\dots,n\},$$

then it follows that $f^{n+1}(x) > 1$ for all $x \in A_n$.

Now we put

$$A = [0,1] \setminus \left(\bigcup_{n=0}^{\infty} A_n \right) = \bigcap_{n=0}^{\infty} ([0,1] \setminus A_n). \tag{3.6}$$

Obviously $[0,1]\setminus A_0$ consists of two closed intervals and $[0,1]\setminus(A_0 \cap A_1)$ of $4 = 2^2$ closed intervals. In general $[0,1]\setminus\left(\bigcup_{k=0}^{n} A_k\right)$ consists of 2^{n+1} closed intervals which are located symmetrically with respect to $\frac{1}{2}$. Every set $[0,1] \setminus A_n$ is closed. Therefore A is non-empty and closed. From the construction it follows that $f(A) \subseteq A$.

In addition we have the following

Theorem 3.9. *The set A defined by (3.6) is a Cantor set, i.e., it is closed, totally unconnected (does not contain intervals) and perfect (every point of A is an accumulation point), if $\mu > 2 + \sqrt{5}$.*

Proof. The closedness of A has been shown already.

$f(x) > 1$ for all $x \in A_0$ implies that

$$A_0 = \Bigl(\underbrace{\frac{1}{2} - \frac{1}{2}\sqrt{1 - \frac{4}{\mu}}}_{x_1}, \underbrace{\frac{1}{2} + \frac{1}{2}\sqrt{1 - \frac{4}{\mu}}}_{x_2}\Bigr).$$

Further we obtain

$$f'(x) \geq f'(x_1) = \mu(1 - 2x_1) = \mu\sqrt{1 - \frac{4}{\mu}} \qquad \text{for all} \quad x \in [0, x_1]$$

and

$$f'(x) \leq f'(x_2) = \mu(1 - 2x_2) = -\mu\sqrt{1 - \frac{4}{\mu}} \qquad \text{for all} \quad x \in [0, x_2].$$

From $\mu > 2 + \sqrt{5}$ it follows that $\mu^2 - 4\mu - 1 > 0$ and further

$$\mu^2\left(1 - \frac{4}{\mu}\right) > 1 \quad \Longrightarrow \quad \mu\sqrt{1 - \frac{4}{\mu}} > 1.$$

This implies

$$|f'(x)| > 1 \qquad \text{for all} \quad x \in [0,1] \setminus A_0$$

and further the existence of some $\lambda > 1$ such that

$$|f'(x)| > \lambda \qquad \text{for all} \quad x \in A \subseteq [0,1] \setminus A_0.$$

The chain rule implies for every $n \in \mathbb{N}$

$$|(f^n)'(x)| > \lambda^n \qquad \text{for all} \quad x \in A.$$

Now let $x, y \in A$ with $x \neq y$ and $[x,y] \subseteq A$ be given. Then it follows for all $n \in \mathbb{N}$ that

$$|(f^n)'(\alpha)| > \lambda^n \qquad \text{for all} \quad \alpha \in [x,y].$$

Choose $n \in \mathbb{N}$ such that $\lambda^n(y-x) > 1$. Then the mean value theorem implies

$$|f^n(y) - f^n(x)| \geq \lambda^n(y-x) > 1$$

and hence $f^n(x) \notin [0,1]$ or $f^n(y) \notin [0,1]$ which contradicts $x, y \in A$.

Therefore A does not contain an interval.

In order to show the perfectness of A we at first observe that every end point of an interval in A_k, $k \in \mathbb{N}_0$, is mapped into 0 after finitely many steps, i.e., belongs to A. Now let $p \in A$ be an isolated point of A. Then there is in every neighbourhood of p a point $x \neq p$ with $f^{k+1}(x) > 1$ for some $k \in \mathbb{N}_0$, hence $x \in A_k$.

Now we distinguish two cases: Either there is a sequence of such points x which are all end points of intervals in some A_k and hence belong to A. This would be a contradiction to the isolation of p. Or every sequence of such points x is mapped elementwise after finitely many steps into the exterior of $[0,1]$. We can even assume that for every element x_k of the sequence there exists some $n \in \mathbb{N}$ with $f^n(x_k) < 0$ so that f^n attains a maximum in p which implies $(f^n)'(p) = 0$. By the chain rule it follows that $f'\left(f^j(p)\right) = 0$ for some $j < n$ and hence $f^j(p) = \frac{1}{2}$ which implies $f^{j+1}(p) > 1$ and $f^n(p) \to \infty$ for $n \to \infty$. This is a contradiction to $p \in A$. Hence the set A does not contain isolated points which completes the proof. □

Remark. *Theorem 3.9 also holds true for $\mu \in (4, 2+\sqrt{5}]$. However, the proof is more complicated.*

For every $x \notin A$ we have $\lim_{n\to\infty} f^n(x) = -\infty$ and for all $x \in A$ it follows that

$$f^n(x) \in A \qquad \text{for all} \quad n \in \mathbb{N}_0.$$

Question: Is $f : A \to A$ chaotic, if $\mu > 2 + \sqrt{5}$?

In order to answer this question in the affirmation sense we show that the mapping $f : A \to A$ and the shift-mapping $\sigma : \Sigma \to \Sigma$ defined by (3.3) are topologically conjugated so that the Corollary 3.8 implies that f is chaotic.

At first we define a mapping $S : A \to \Sigma$ by $S(x) = (s_0, s_1, s_2, \ldots)$ where

$$s_j = \begin{cases} 0, & \text{if} \quad f^j(x) \in I_0 \\ 1, & \text{if} \quad f^j(x) \in I_1, \end{cases} \tag{3.7}$$

and I_0, I_1 are the two intervals with $I_0 \cup I_1 = [0,1] \backslash (A_0 = \{x \in [0,1] \mid f(x) > 1\})$.

For this mapping we prove the

Theorem 3.10. *If* $\mu > 2 + \sqrt{5}$, *then* $S : A \to \Sigma$ *is a homeomorphism.*

Proof. At first we prove that S is injective. Let $x, y \in A$ with $x \neq y$ be given. Assumption: $S(x) = S(y)$. Then, for every $n \in \mathbb{N}_0$, it follows that

$$(f^n(x) \in I_0 \quad \text{and} \quad f^n(y) \in I_0) \qquad \text{or} \qquad (f^n(x) \in I_1 \quad \text{and} \quad f^n(y) \in I_1)\,.$$

This implies $[x, y] \subseteq A$ or $[y, x] \subseteq A$ which contradicts the fact that by Theorem 3.9 A is totally disconnected.

In order to show that S is surjective we define, for every closed interval $J \subseteq [0,1]$, the n−th counterimage by

$$f^{-n}(J) = \{x \in [0,1] \mid f^n(x) \in J\}\,.$$

The counterimage $f^{-1}(J)$ consists of two intervals, one in I_0 and another one in I_1 as being shown in the following picture:

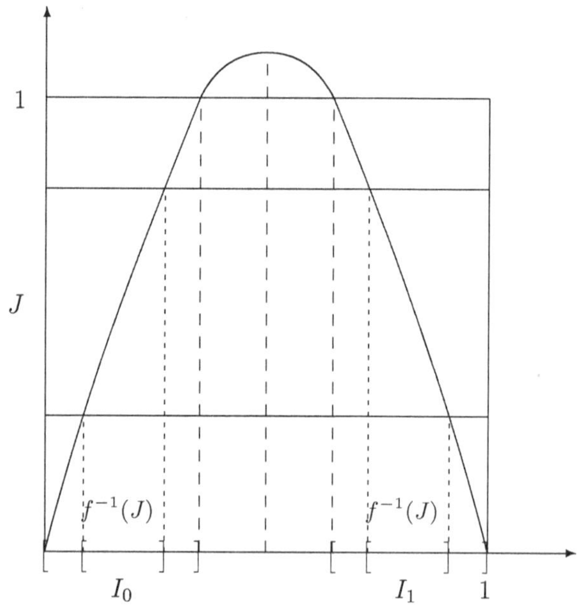

Now let $s = (s_0, s_1, s_2, \ldots) \in \Sigma$ be given. Then we define for every $n \in \mathbb{N}_0$

$$\begin{aligned} I_{s_0 s_1 \ldots s_n} &= \{x \in [0,1] \mid x \in I_{s_0},\ f(x) \in I_{s_1}, \ldots,\ f^n(x) \in I_{s_n}\} \\ &= I_{s_0} \cap f^{-1}(I_{s_1}) \cap \ldots \cap f^{-n}(I_{s_n}) = I_{s_0} \cap f^{-1}(I_{s_1 s_2 \ldots s_n}). \end{aligned} \tag{3.8}$$

According to the above remark $f^{-1}(I_{s_1 s_2 \ldots s_n})$ consists of two closed intervals, one in I_0 and another one in I_1. Therefore $I_{s_0} \cap f^{-1}(I_{s_1 s_2 \ldots s_n})$ is a closed interval in I_{s_0}. Further we have

$$I_{s_0 \ldots s_n} = I_{s_0 \ldots s_{n-1}} \cap f^{-n}(I_{s_n}) \subseteq I_{s_0 \ldots s_{n-1}}.$$

This implies that $\bigcap\limits_{n \in \mathbb{N}_0} I_{s_0 s_1 \ldots s_n}$ is non-empty. Let $x \in \bigcap\limits_{n \in \mathbb{N}_0} I_{s_0 s_1 \ldots s_n}$. Then it follows that

$$x \in I_{s_0},\ f(x) \in I_{s_1},\ f^2(x) \in I_{s_2}, \ldots,$$

hence $S(x) = s$, which shows the surjectivity of S. Further we have

$$\bigcap_{n \in \mathbb{N}_0} I_{s_0 s_1 \ldots s_n} = \{x\}, \quad \text{since } S \text{ is injective.}$$

In order to show the continuity of S we choose an arbitrary $x \in A$ and put $(s_0, s_1, \ldots) = S(x)$. Let $\varepsilon > 0$ be given. Then we choose $n \in \mathbb{N}_0$ such that $\frac{1}{2^n} \leq \varepsilon$. Now consider the intervals $I_{t_0 t_1 \ldots t_n}$ defined by (3.8) for all 2^{n+1} combinations of $t_0, \ldots, t_n$. These are all different from each other and $I_{s_0 s_1 \ldots s_n}$ is one of these. Now choose $\delta > 0$ such that

$$|x - y| < \delta \quad \text{and} \quad y \in A \quad \Longrightarrow \quad y \in I_{s_0 s_1 \ldots s_n}.$$

Then it follows that

$$S(x)_i = S(y)_i \qquad \text{for} \quad i = 0, \ldots, n$$

and hence by Lemma 3.2

$$d(S(x), S(y)) \leq \frac{1}{2^{n+1}} \leq \varepsilon$$

which proves the continuity of S.

In order to show the continuity of the inverse mapping $S^{-1} : \Sigma \to A$ we show that S maps closed subsets of A into closed subsets of Σ. Let $B \subseteq A$ be closed and non-empty. Then B is compact, since A is compact as a closed subset of $[0,1]$ which is compact. Since S is continuous $S(B)$ is compact in Σ and therefore closed.

This completes the proof of Theorem 3.10. □

After these preparations we come to

Theorem 3.11. *The mappings $f : A \to A$ and $\sigma : \Sigma \to \Sigma$ are topologically conjugated, if $\mu > 2 + \sqrt{5}$. In particular we have*

$$S \circ f = \sigma \circ S$$

where $S : A \to \Sigma$ is the homeomorphism defined by (3.7).

Proof. From the proof of Theorem 3.10 we obtain that every $x \in A$ can be represented uniquely in the form

$$\{x\} = \bigcap_{n \in \mathbb{N}_0} I_{s_0 s_1 \ldots s_n} \qquad \text{where} \quad (s_0, s_1, \ldots) = S(x).$$

From (3.8) it follows because of $f(I_{s_0}) = [0,1]$ that

$$f\left(I_{s_0 s_1 \ldots s_n}\right) = I_{s_1} \cap f^{-1}\left(I_{s_2}\right) \cap \ldots \cap f^{-n+1}\left(I_{s_1 s_2 \ldots s_n}\right).$$

Therefore it follows that

$$S(f(x)) = S \circ f\left(\bigcap_{n \in \mathbb{N}_0} I_{s_0 s_1 \ldots s_n}\right) = S\left(\bigcap_{n \in \mathbb{N}} I_{s_1 s_2 \ldots s_n}\right) = (s_1, s_2, \ldots) = \sigma(S(x))$$

which completes the proof. □

Corollary 3.8, Theorem 3.6 and Theorem 3.11 now imply

Theorem 3.12. *The mapping $f : A \to A$ defined by (3.5) with A defined by (3.6) is chaotic, if $\mu > 2 + \sqrt{5}$.*

Next we consider the mapping f (3.5) for $\mu = 4$, i.e,

$$f(x) = 4x(1-x), \qquad x \in \mathbb{R}.$$

Obviously f maps the interval $[0,1]$ into itself. Since f is continuous, it defines a time-discrete dynamical system on $X = [0,1]$.

In order to show that f is chaotic we at first consider the space Σ_2 of twofold-infinite $0-1-$sequences, i.e.,

$$\Sigma_2 = \left\{ s = (s_j)_{j \in \mathbb{Z}} \mid s_j = 0 \ \text{ or } \ 1 \right\}.$$

If we equip Σ_2 with the metric

$$d(s,t) = \sum_{j=-\infty}^{\infty} \frac{1}{2^{|j|}} |s_j - t_j|, \qquad s, t \in \Sigma_2,$$

then Σ_2 becomes a compact metric space which is homeomorphic to the interval $[0,1]$ where the homeomorphism is given by the mapping

$$s = (s_j)_{j \in \mathbb{Z}} \to \sum_{j \in \mathbb{Z}} \frac{1}{2^{|j|}} s_j, \qquad s \in \Sigma_2.$$

Lemma 3.13. *Given $s, t \in \Sigma_2$. Then the following two statements are true:*

(a) $s_i = t_i$ *for* $i = -n, \ldots, n \implies d(s,t) \le \frac{1}{2^{n-1}}$.

(b) $d(s,t) < \frac{1}{2^n} \implies s_i = t_i$ *for* $i = -n, \ldots, n$.

The proof of this lemma is the same as that for sequences in Σ.

We again define a shift-mapping $\sigma : \Sigma_2 \to \Sigma_2$ by

$$\sigma(s)_j = s_{j+1} \qquad \text{for all} \quad j \in \mathbb{Z} \text{ and all } s \in \Sigma_2.$$

This mapping is even a homeomorphism, i.e., bi-unique, surjective and continuous together with its inverse mapping $\sigma^{-1} : \Sigma_2 \to \Sigma_2$ which is defined by

$$\sigma^{-1}(t) = t_{j-1} \qquad \text{for all} \quad j \in \mathbb{Z} \text{ and all } t \in \Sigma_2.$$

For every $n \in \mathbb{N}_0$ there are 2^{2n+1} different points $s \in \Sigma_2$ of the form

$$(\ldots, s_n, s_{-n+1}, \ldots, s_0, \ldots, s_{n-1}, s_n, \ldots)$$

for which it follows that $\sigma^{2n+1}(s) = s$, i.e., which are period points. Similar to Theorem 3.4 and Theorem 3.5 one can prove

Assertion 1: The set $\mathrm{Per}(\sigma)$ of period points is dense in Σ_2.

Assertion 2: There is some $s \in \Sigma_2$ such that the orbit $(\sigma^n(s))_{n \in \mathbb{N}_0}$ is dense in Σ_2.

As a consequence we obtain the statement (see Theorem 3.6):

The shift-mapping $\sigma : \Sigma_2 \to \Sigma_2$ is topologically transitive and hence chaotic.

Since Assertion 1 and 2 are also true for the inverse mapping $\sigma^{-1} : \Sigma_2 \to \Sigma_2$, it follows that σ^{-1} is also chaotic.

Next we consider the mapping $g : S^1 \to S^1$ of the unit circle $S^1 = \{e^{i\alpha} \mid \alpha \in \mathbb{R}\}$ into itself which is defined by

$$g\left(e^{i\alpha}\right) = e^{2i\alpha}, \qquad \alpha \in \mathbb{R}.$$

Now let $h : \Sigma_2 \to S^1$ be defined by

$$h(s) = e^{i\alpha} \qquad \text{where} \quad \alpha = \sum_{j \in \mathbb{Z}} \frac{s_j}{2^j}.$$

Then h is continuous and surjective. Further we have

$$h \circ \sigma^{-1} = g \circ h.$$

Theorem 3.7 implies then that g is also chaotic.

If we define a mapping $h_1 : S^1 \to [-1,1]$ by

$$h_1\left(\mathrm{e}^{i\alpha}\right) = \cos\alpha, \qquad \alpha \in \mathbb{R}$$

and a mapping $h_2[-1,1] \to [0,1]$ by

$$h_2(x) = \frac{1}{2}(1-x), \qquad x \in [-1,1],$$

then $h_2 \circ h_1 : S^1 \to [0,1]$ is continuous and surjective.

Further we have

$$h_2 \circ h_1 \circ g\left(\mathrm{e}^{i\alpha}\right) = \frac{1}{2}(1-\cos 2\alpha) = \sin^2\alpha$$

and

$$\begin{aligned} f \circ h_2 \circ h_1\left(\mathrm{e}^{i\alpha}\right) &= f \circ h_2(\cos\alpha) = f\left(\frac{1}{2}(1-\cos\alpha)\right) \\ &= 2\,(1-\cos\alpha)\left(1-\frac{1}{2}(1-\cos\alpha)\right) \\ &= 1-\cos^2\alpha = \sin^2\alpha \qquad \text{for} \quad \alpha \in \mathbb{R}, \end{aligned}$$

hence $h_2 \circ h_1 \circ g = f \circ h_2 \circ h_1$.

Again Theorem 3.7 implies that f is chaotic.

3.3 The Topological Entropy as a Measure for Chaos

3.3.1 Definition and Invariance

The following representation is based on the paper [28] by A. Mielke.

Let X be a compact metric space whose metric is denoted by d and let $f : X \to X$ be a continuous mapping.

Definition. *Choose $n \in \mathbb{N}$ and $\varepsilon > 0$ arbitrarily. A non-empty subset $M \subseteq X$ is called $(n, \varepsilon)-$separated with respect to f, if for every pair $x \neq y$ in M there exists a number $m \in \{1, 2, \ldots, n\}$ with*

$$d\left(f^m(x), f^m(y)\right) > \varepsilon.$$

Assertion: If $M \subseteq X$ is $(n, \varepsilon)-$ separated with respect to f, then M consists at most of finitely many elements.

Proof. If M consisted of infinitely many elements, then there would exist an accumulation point $\widehat{x} \in X$ of M and a sequence $(x_i)_{i \in \mathbb{N}}$ of different points in M with $x_i \neq \widehat{x}$ for all $i \in \mathbb{N}$ and $\lim\limits_{i \to \infty} d\left(x_i, \widehat{x}\right) = 0$ which in particular implies $\lim\limits_{i \to \infty} d\left(x_i, x_{i+1}\right) = 0$. Because of the continuity of f (which implies uniform continuity) there exists, for all $m \in \{1, 2, \ldots, n\}$, some $i(\varepsilon) \in \mathbb{N}$ with

$$d\left(f^m\left(x_i\right), f^m\left(x_{i+1}\right)\right) \leq \varepsilon \qquad \text{for all} \quad m \in \{1, \ldots, n\} \text{ and all } i \geq i(\varepsilon).$$

This contradicts the $(n, \varepsilon)-$separatedness of M with respect to f which completes the proof. □

For every $n \in \mathbb{N}$ and $\varepsilon > 0$ we put

$$s(f, n, \varepsilon) = \sup\{\#M \mid M \subseteq X \text{ is } (n, \varepsilon) - \text{separated with respect to } f\}$$

where $\#$ denotes the number of elements of M and define

$$h(f, \varepsilon) = \limsup_{n \to \infty} \frac{1}{n} \ln(s(f, n, \varepsilon)).$$

Let $\varepsilon_2 > \varepsilon_1 > 0$. Then every $(n, \varepsilon_2)-$separated subset of X with respect to f is also $(n, \varepsilon_1)-$separated which implies

$$s\left(f, n, \varepsilon_1\right) \geq s\left(f, n, \varepsilon_2\right) \quad \Longrightarrow \quad h\left(f, \varepsilon_1\right) \geq h\left(f, \varepsilon_2\right).$$

The topological entropy of f is then defined by

$$h(f) = \lim_{\varepsilon \to 0+} h(f, \varepsilon) = \sup_{0 < \varepsilon < \varepsilon_0} h(f, \varepsilon) \qquad \text{for all} \quad \varepsilon_0 > 0.$$

Theorem 3.14. *Let X_1 and X_2 be two compact metric spaces with metrics d_1 and d_2, respectively. Further let $f_i : X_i \to X_i$ for $i = 1, 2$ be two continuous mappings. If there is an injective continuous mapping $g : X_1 \to X_2$ with $f_2 \circ g = g \circ f_1$, then it follows that $h(f_1) \le h(f_2)$.*

Proof. Since $g : X_1 \to X_1$ is bi-unique and continuous, the inverse mapping $g^{-1} : g(X_1) \to X_1$ is a homeomorphism and hence uniformly continuous. For every $\varepsilon > 0$ therefore exists a $\delta(\varepsilon) > 0$ with

$$\lim_{\varepsilon \to 0+} \delta(\varepsilon) = 0 \qquad \text{and} \qquad d_2(g(x), g(y)) \le \delta(\varepsilon) \implies d_1(x, y) \le \varepsilon.$$

The latter implication is equivalent to

$$d_1(x, y) > \varepsilon \implies d_2(g(x), g(y)) > \delta(\varepsilon).$$

Now let $M_1 \subseteq X_1$ be $(n, \varepsilon)-$separated with respect to f_1. Then for $x \ne y$ in M_1 there is some $m \in \{1, \dots, n\}$ with $d_1(f_1^m(x), f_1^m(y)) > \varepsilon$. This implies

$$d_2(f_2^m(g(x)), f_2^m(g(y))) = d_2(g(f_1^m(x)), g(f_1^m(y))) > \delta(\varepsilon).$$

Because of $g(x) \ne g(y) \Leftrightarrow x \ne y$ it follows that $g(M_1) \subseteq g(X_1)$ is $(n, \delta(\varepsilon))-$separated with respect to f_2. This implies

$$s(f_1, n, \varepsilon) \le s(f_2, n, \delta(\varepsilon))$$

and further $h(f_1, \varepsilon) \le h(f_2, \delta(\varepsilon))$ which finally implies $h(f_1) \le h(f_2)$.

This completes the proof. □

Theorem 3.15. *We assume that under the assumption of Theorem 3.14 there exists a surjective continuous mapping $g : X_1 \to X_2$ with $f_2 \circ g = g \circ f_1$. Then it follows that $h(f_2) \le h(f_1)$.*

Proof. Since $g : X_1 \to X_2$ is uniformly continuous, there exists for every $\varepsilon > 0$ some $\delta(\varepsilon)$ such that

$$\lim_{\varepsilon \to 0+} \delta(\varepsilon) = 0 \qquad \text{and} \qquad d_1(x, y) \le \delta(\varepsilon) \implies d_2(g(x), g(y)) \le \varepsilon.$$

The latter implication is equivalent to

$$d_2(g(x), g(y)) > \varepsilon \implies d_1(x, y) > \delta(\varepsilon).$$

Now let $M_2 \subseteq X_2$ be $(n, \varepsilon)-$separated with respect to f_2. Then for $u \ne v$ in M_2 there exists some $m \in \{1, \dots, n\}$ with $d_2(f_2^m(u), f_2^m(v)) > \varepsilon$.

Let $x, y \in X_1$ with $u = g(x)$ and $v = g(y)$ be given. Then it follows that $x \ne y$ and

$$d_2(g(f_1^m(x)), g(f_1^m(y))) = d_2(f_2^m(u), f_2^m(v)) > \varepsilon.$$

This implies $d_1\left(f_1^m(x), f_1^m(y)\right) > \delta(\varepsilon)$ for some $m \in \{1, \ldots, n\}$.

Therefore $g^{-1}(M_2) \subseteq X_2$ is $(n, \delta(\varepsilon))$–separated with respect to f. Because of $\#M_2 \leq \#g^{-1}(M_2)$ we obtain

$$s(f_2, n, \varepsilon) \leq s(f_1, n, \delta(\varepsilon))$$

and further $h(f_2, \varepsilon) \leq h(f_1, \delta(\varepsilon))$ which finally implies $h(f_2) \leq h(f_1)$. □

As a consequence of Theorem 3.14 and Theorem 3.15 we obtain the

Theorem 3.16. *Let the continuous mappings $f_1 : X_1 \to X_1$ and $f_2 : X_2 \to X_2$ of two compact metric spaces X_1 and X_2 be topologically conjugated.*

Then it follows that $h(f_1) = h(f_2)$.

This property of the topological entropy is called topological invariance.

Definition. *A continuous mapping $f : X \to X$ of a compact metric space X into itself is called disorder-chaotic, if the topological entropy $h(f)$ of f is positive ($h(f) \geq 0$ follows from the definition).*

3.3.2 The Topological Entropy of the Shift-Mapping

In Section 3.1 we have considered the space Σ of the $0-1$–sequences $s = (s_0, s_1, s_2, \ldots)$ with $s_j = 1$ or $s_j = 0$ for every $j \in \mathbb{N}_0$ endowed with the metric

$$d(s,t) = \sum_{i \in \mathbb{N}_0} \frac{|s_i - t_i|}{2^i} \qquad \text{for} \quad s, t \in \Sigma.$$

This space is compact, since by virtue of the mapping

$$s = (s_0, s_1, s_2, \ldots) \longmapsto d(s) = \sum_{i \in \mathbb{N}_0} \frac{s_i}{2^i},$$

it turns out to be homeomorphic to the compact interval $[0, 2]$. The shift-mapping $\sigma : \Sigma \to \Sigma$ is defined by

$$\sigma(s_0, s_1, s_2, \ldots) = (s_1, s_2, s_3, \ldots)$$

and is continuous according to Theorem 3.3. We even have

$$d(\sigma(s), \sigma(t)) \leq 2d(s,t) \qquad \text{for} \quad s, t \in \Sigma$$

which shows that σ is Lipschitz-continuous.

In the following we look for the topological entropy $h(\sigma)$ of σ. For that purpose we consider for every $n \in \mathbb{N}$ the subset

$$\Sigma_n = \{s \in \Sigma \mid s_{k+n} = s_k \quad \text{for all} \quad k \in \mathbb{N}_0\} \quad \text{of} \quad \Sigma.$$

Let $s, t \in \Sigma_n$ be given with $s \neq t$. Then it follows for every $m \in \{1, \ldots, n\}$ that

$$d\left(\sigma^m(s), \sigma^m(t)\right) = \sum_{k \in \mathbb{N}_0} \frac{1}{2^k} |s_{k+m} - t_{k+m}|,$$

hence,

$$d\left(\sigma^m(s), \sigma^m(t)\right) \geq 1, \qquad \text{if} \quad s_m \neq t_m.$$

Therefore Σ_n is $(n, \varepsilon)-$separated with respect to σ for every $\varepsilon \in (0,1)$. This implies

$$2^n = \#\Sigma_n \leq s(\sigma, n, \varepsilon) \qquad \text{for all} \quad \varepsilon \in (0,1)$$

and further

$$\ln 2 \leq \frac{1}{n} \limsup_{n \to \infty} s(\sigma, n, \varepsilon) = h(\sigma, \varepsilon) \qquad \text{for all} \quad \varepsilon \in (0,1)$$

which implies

$$\ln 2 \leq h(\sigma)$$

and shows that σ is disorder-chaotic.

Next we will determine an upper bound for $h(\sigma)$. For that purpose we start with the implication

$$\begin{aligned} d\left(\sigma^m(s), \sigma^m(t)\right) > 2^{1-N} \quad &\text{for some} \quad m \in \{1, \ldots, n\} \\ \Longrightarrow \qquad d(s,t) > 2^{1-(N+m)} \quad &\text{for every pair} \quad n, N \in \mathbb{N}. \end{aligned}$$

If we choose $\varepsilon = 2^{1-N}$ and assume that $M \subseteq \Sigma$ is $(n, \varepsilon)-$separated with respect to σ, then, for every pair $s, t \in M$, there is some $m \in \{1, \ldots, n\}$ with $d\left(\sigma^m(s), \sigma^m(t)\right) > \varepsilon$ which implies $d(s,t) > 2^{1-(N+m)}$.

Therefore Lemma 3.2 implies

$$s_k \neq t_k \qquad \text{for some} \quad k \in \{0, \ldots, N+n-1\}.$$

From this it follows that

$$s\left(\sigma, n, \frac{1}{2^{N-1}}\right) \leq \#\Sigma_{N+n} = 2^{N+n}$$

and it is

$$\frac{1}{n} \ln\left(s\left(\sigma, n, \frac{1}{2^{N-1}}\right)\right) \leq \left(1 + \frac{N}{n}\right) \ln 2.$$

Finally this implies

$$h\left(\sigma, \frac{1}{2^{N-1}}\right) = \limsup_{n \to \infty} \frac{1}{n} \ln\left(s\left(\sigma, n, \frac{1}{2^{N-1}}\right)\right) \leq \ln 2 \qquad \text{for all} \quad N \in \mathbb{N}$$

and therefore $h(\sigma) \leq \ln 2$ which, with the above result, implies $h(\sigma) = \ln 2$.

3.3.3 Disorder-Chaos for One-Dimensional Mappings

In order to show disorder-chaos for one-dimensional mappings it is useful to determine the topological entropy of suitable subspaces of Σ. For that purpose we introduce $2 \times 2-$matrices

$$A = \begin{pmatrix} a_{00} & a_{01} \\ a_{10} & a_{11} \end{pmatrix}$$

with $a_{ij} = 0$ or 1 for $i, j = 0, 1$. Then we define, for a given such matrix, a subspace Σ^A of Σ by

$$\Sigma^A = \left\{ (s_i)_{i \in \mathbb{N}_0} \in \{0,1\}^{\mathbb{N}} \mid a_{s_i s_{i+1}} = 1 \quad \text{for all} \quad i \in \mathbb{N}_0 \right\}.$$

<u>Assertion:</u> Σ^A is closed and therefore compact.

Proof. Let $\left(s^k\right)_{k \in \mathbb{N}}$ be a sequence in Σ^A with $s^k \to t$ for some $t \in \Sigma$.

Assumption: $t \notin \Sigma^A$. Then there is a smallest $i_0 \in \mathbb{N}_0$ with $a_{t_{i_0} t_{i_0+1}} = 0$. Because of $s^k \to t$ there is some $k_0 \in \mathbb{N}$ with

$$d\left(s^k, t\right) < \frac{1}{2^{i_0+1}} \qquad \text{for all} \quad k \geq k_0.$$

By Lemma 3.2 it follows that

$$s_i^k = t_i \qquad \text{for} \quad i = 0, \ldots, i_0 + 1.$$

Because of $s^k \in \Sigma^A$ for all $k \in \mathbb{N}$ this implies $a_{t_{i_0} t_{i_0+1}} = 1$ which leads to a contradiction and concludes the proof. □

Obviously we have $\sigma\left(\Sigma^A\right) \subseteq \Sigma^A$ so that σ also defines a dynamical system. We denote the restriction of σ to Σ^A by σ_A and call this mapping a subshift.

<u>Question.</u> How does one calculate the topological entropy of σ_A?
From the above considerations we at first deduce

$$|\Sigma_n^A| \leq s\left(\sigma_A, n, \varepsilon\right) \qquad \text{for all} \quad \varepsilon \in (0,1) \quad \text{and all} \quad n \in \mathbb{N}$$

and

$$s\left(\sigma_A, n, \frac{1}{2^{N-1}}\right) \leq |\Sigma_{N+n}^A| \qquad \text{for all} \quad n, N \in \mathbb{N},$$

where for every $n \in \mathbb{N}$

$$\begin{aligned} \Sigma_n^A = \{&(s_0, s_1, \ldots, s_{n-1}, s_0, s_1, \ldots, s_{n-1}, \ldots) \in \Sigma \mid a_{s_0 s_1} = 1, \ldots, \\ &a_{s_{n-2} s_{n-1}} = 1\}. \end{aligned}$$

Let $e = (1,1)^T$ and for every $n \in \mathbb{N}$ let

$$a(n) = \begin{pmatrix} a_0(n) \\ a_1(n) \end{pmatrix} = A^{n-1} e.$$

<u>Assertion:</u> For every $n \in \mathbb{N}$ it is true that

$$|\Sigma_n^A| = \|A^{n-1} e\|_1 = a_0(n) + a_1(n).$$

Proof. Let $n = 1$. Then we have

$$\Sigma_1^A = \{(s_0, s_1, \ldots) \in \Sigma \mid s_0 = 0 \text{ or } s_0 = 1\}$$

and

$$|\Sigma_1^A| = 2 = a_0(1) + a_1(1).$$

Let $n = 2$. Then

$$|\Sigma_2^A| = a_{00} + a_{01} + a_{10} + a_{11} = \|Ae\|_1 = a_0(2) + a_1(2).$$

Let $n = 3$. Then

$$|\Sigma_3^A| = a_{00}^2 + a_{01}a_{10} + a_{10}a_{00} + a_{11}a_{10} + a_{00}a_{01} + a_{01}a_{11} + a_{10}a_{01} + a_{11}^2.$$

Now we have

$$A^2 = \begin{pmatrix} a_{00}^2 + a_{01}a_{10} & a_{00}a_{01} + a_{01}a_{11} \\ a_{10}a_{00} + a_{11}a_{10} & a_{10}a_{01} + a_{11}^2 \end{pmatrix}$$

which implies

$$|\Sigma_3^A| = \|A^2 e\|_1 = a_0(3) + a_1(3).$$

In general for $n \geq 2$ we have

$$\begin{aligned} |\Sigma_n^A| = {} & \text{sum of all products } a_{s_0 s_1} a_{s_1 s_2} \cdots a_{s_{n-2} s_{n-1}} \\ & \text{with } s_i = 0 \text{ or } 1 \text{ for } i = 0, \ldots, n-1 \\ = {} & \|A^{n-1} e\|_1 = a_0(n) + a_1(n). \end{aligned}$$

□

The first of the above two statements therefore implies, for all $\varepsilon \in (0,1)$ and all $n \in \mathbb{N}$,

$$\frac{1}{n} \ln\left(a_0(n) + a_1(n)\right) \leq \frac{1}{n} \ln\left(s\left(\sigma_A, n, \varepsilon\right)\right)$$

and from the second we infer for all $n, N \in \mathbb{N}$ that

$$\frac{1}{n} \ln\left(s\left(\sigma_A, n, \frac{1}{2^{N-1}}\right)\right) \leq \frac{1}{n} \ln\left(a_0(N+n) + a_1(N+n)\right).$$

Now we have

$$\begin{pmatrix} a_0(N+n) \\ a_1(N+n) \end{pmatrix} = A^N \begin{pmatrix} a_0(n) \\ a_1(n) \end{pmatrix} = \begin{pmatrix} b_{00}^{(N)} & b_{01}^{(N)} \\ b_{10}^{(N)} & b_{11}^{(N)} \end{pmatrix} \begin{pmatrix} a_0(n) \\ a_1(n) \end{pmatrix},$$

hence

$$\begin{aligned} a_0(N+n) + a_1(N+n) &= \left(b_{00}^{(N)} + b_{10}^{(N)}\right) a_0(n) + \left(b_{01}^{(N)} + b_{11}^{(N)}\right) a_1(n) \\ &\leq \max\left(b_{00}^{(N)} + b_{10}^{(N)}, b_{01}^{(N)} + b_{11}^{(N)}\right)\left(a_0(n) + a_1(n)\right) \\ &= \gamma_N \left(a_0(n) + a_1(n)\right). \end{aligned}$$

This implies

$$\frac{1}{n}\ln\left(a_0(N+n)+a_1(N+n)\right) \le \frac{1}{n}\ln\left(a_0(n)+a_1(n)\right)+\frac{1}{n}\ln\gamma_N$$

and further

$$\limsup_{n\to\infty}\frac{1}{n}s\left(\sigma_A,n,\frac{1}{2^{N-1}}\right) \le \limsup_{n\to\infty}\frac{1}{n}\ln\left(a_0(n)+a_1(n)\right) \quad \text{for all} \quad N\in\mathbb{N}.$$

On the other hand we have

$$\limsup_{n\to\infty}\frac{1}{n}\ln\left(a_0(n)+a_1(n)\right) \le \limsup_{n\to\infty}\frac{1}{n}\ln\left(s\left(\sigma_A,n,\varepsilon\right)\right)$$

for all $\varepsilon\in(0,1)$.

Therefore we obtain for all $N\ge 2$

$$h\left(\sigma_A,\frac{1}{2^{N-1}}\right)=\limsup_{n\to\infty}\frac{1}{n}\ln\left(a_0(n)+a_1(n)\right)$$

and therefore

$$h\left(\sigma_A\right)=\limsup_{n\to\infty}\frac{1}{n}\ln\left(a_0(n)+a_1(n)\right).$$

Without proof we use the statement

$$\lim_{n\to\infty}\frac{1}{n}\ln\left(\|A^{n-1}\mathrm{e}\|_1\right)=\ln(\lambda(A))$$

where $\lambda(A)$ is the largest eigenvalue of A.

As a result we obtain the

Theorem 3.17. *The topological entropy of* σ_A *is given by*

$$h\left(\sigma_A\right)=\ln(\lambda(A))$$

where $\lambda(A)$ *is the largest eigenvalue of* A.

Example. $A=\begin{pmatrix}0&1\\1&1\end{pmatrix}$.

In this case we have $\lambda(A)=\frac{1}{2}(1+\sqrt{5})$ and therefore

$$h\left(\sigma_A\right)=\ln\left(\frac{1}{2}(1+\sqrt{5})\right)>0.$$

We will now apply this result in order to show that a continuous mapping $f : \mathbb{R} \to \mathbb{R}$ which has a period point of $f-$period 3 is disorder-chaotic on a suitable subset $J \subseteq \mathbb{R}$.

Let there be given three points $a, b, c \in \mathbb{R}$ with $a < b < c$ and $f(a) = b,\ f(b) = c$ and $f(c) = a$. We put $I_0 = [a, b]$ and $I_1 = [b, c]$. Then it follows that

$$I_1 \subseteq f(I_0) \quad \text{and} \quad I_0 \cup I_1 \subseteq f(I_1), \qquad \text{however in general} \quad I_0 \not\subseteq f(I_0).$$

We even assume that $I_0 \cap f(I_0) = \{b\}$. If we define for $i, j = 0, 1$

$$a_{ij} = \begin{cases} 0, & \text{if} \quad I_j \not\subseteq f(I_i), \\ 1, & \text{if} \quad I_i \subseteq f(I_j), \end{cases}$$

then we obtain the matrix

$$A = \begin{pmatrix} 0 & 1 \\ 1 & 1 \end{pmatrix}.$$

If we define for every $n \in \mathbb{N}_0$ the set

$$J_n = \{x \in J_0 \mid f^n(x) \in J_0\} \qquad \text{where} \quad J_0 = [a, c] = I_0 \cup I_1,$$

then $J = \bigcap_{n \in \mathbb{N}_0} J_n$ is a non-empty compact subset of $\mathbb{R}$ with $f(J) \subseteq J$, since every J_n is a counterimage of a continuous mapping of J_0 which is a compact interval. If J were empty, then already a finite intersection $\bigcap_{n \in \mathbb{N}} J_n$ were empty which is not the case. Now let $x \in J$ be given. Then it follows that $f^n(f(x)) \in J_0$, i.e., $f(x) \in J$ and therefore $f(J) \subseteq J$.

Finally we define the mapping $S : J \to \Sigma$ by $S(x) = (s_0, s_1, s_2, \ldots)$ where

$$s_j = \begin{cases} 0, & \text{if} \quad f^j(x) \in [a, b), \\ 1, & \text{if} \quad f^j(x) \in [b, c). \end{cases}$$

Obviously it follows that

$$s_j = 0 \quad \Longrightarrow \quad s_{j+1} = 1 \ \text{ and } \ s_j = 1 \quad \Longrightarrow \quad s_{j+1} = 0 \ \text{ or } \ s_{j+1} = 1.$$

This implies $f(J) \subseteq \Sigma^A$.

<u>Assertion 1:</u> The mapping $S : J \to \Sigma^A$ is surjective.

Proof. Let $(s_0, s_1, s_2, \ldots) \in \Sigma^A$ be given. Then it follows that

$$I_{s_1} \subseteq f(I_{s_0}),\ I_{s_2} \subseteq f(I_{s_1}), \ldots, I_{s_n} \subseteq f\left(I_{s_{n-1}}\right) \qquad \text{for every} \quad n \in \mathbb{N}.$$

If we put $A_i = I_{s_i}$ for $i \in \mathbb{N}_0$, then it follows that

$$A_{i+1} \subseteq f(A_i) \qquad \text{for all} \quad i \in \mathbb{N}_0.$$

Then, for every $i \in \mathbb{N}_0$, there exists a closed set

$$B_i \subseteq A_0 = I_{s_0} \quad \text{with} \quad B_{i+1} \subseteq B_i \quad \text{and} \quad f^i\left(B_i\right) = A_i = I_{s_i}.$$

In order to see that we put $B_0 = A_0$ and obtain $f^0\left(B_0\right) = A_0$.

Now assume B_i for some $i \in \mathbb{N}_0$ already to be defined such that

$$B_i \subseteq A_0 \quad \text{and} \quad f^i\left(B_i\right) = A_i.$$

From $A_{i+1} \subseteq f\left(A_i\right)$ it follows that $A_{i+1} \subseteq f^{i+1}\left(B_i\right)$. This implies the existence of a closed set $B_{i+1} \subseteq B_i$ with $f^{i+1}\left(B_{i+1}\right) = A_{i+1}$ (Simply put $B_{i+1} = f^{-i-1}\left(A_{i+1}\right)$).

If we put $B = \bigcap\limits_{n \in \mathbb{N}_0} B_n$ then B is non-empty and compact (by the same arguments as for J) and it follows that

$$f^n(x) \in I_{s_n} \quad \text{for all} \quad x \in B \text{ and all } n \in \mathbb{N}_0.$$

This shows that $B \subseteq J$ and $(s_0, s_1, s_2, \ldots) = S(x)$ for all $x \in B$. □

Assertion 2: The mapping $S : J \to \Sigma^A$ is continuous.

Proof. Let $x \in J$ and $\varepsilon > 0$ be chosen arbitrarily. Then we choose $n \in \mathbb{N}_0$ such that $\frac{1}{2^n} < \varepsilon$. Let $S(x) = (s_0, s_1, s_2, \ldots)$. For this sequence in Σ^A we assume a sequence $B_i \subseteq I_{s_0}$, $i \in \mathbb{N}_0$, to be constructed as in the proof of Assertion 1 so that in particular $x \in B = \bigcap\limits_{i \in \mathbb{N}_0} B_i$ and

$$f^i(x) \in I_{s_i} \quad \text{for all} \quad i \in \mathbb{N}_0.$$

Now let $y \in B_n$ be chosen arbitrarily. Then it follows that

$$f^i(y) \in I_{s_i} \quad \text{for all} \quad i = 0, \ldots, n$$

and therefore

$$S(y)_j = s_j = S(x)_j \quad \text{for} \quad j = 0, \ldots, n.$$

By Lemma 3.2 this implies

$$d(S(x), S(y)) \leq \frac{1}{2^n} < \varepsilon$$

which proves the continuity of S. □

Further we infer from the definition of S that

$$S \circ f = \sigma_A \circ S.$$

From Theorem 3.15 and Theorem 3.17 we now obtain the statement

$$h(f) \geq h\left(\sigma_A\right) = \ln\left(\frac{1}{2}(1 + \sqrt{5})\right).$$

Result. *The above mapping $f : \mathbb{R} \to \mathbb{R}$ is disorder-chaotic on J.*

3.4 Chaos in the Sense of Li and Yorke

In 1975 Tien-Yien Li and James A. Yorke have introduced the following concept of chaos: Given a continuous function $f : J \to J$ where $J \subseteq \mathbb{R}$ is a finite closed interval. This function is called chaotic (in the sense of Li and Yorke), if there exists an uncountable set $S \subseteq J$ without period points with the following properties:

(a) For every pair $p, q \in S$ with $p \neq q$ it follows that

$$\limsup_{n\to\infty} |f^n(p) - f^n(q)| > 0 \tag{3.9}$$

and

$$\liminf_{n\to\infty} |f^n(p) - f^n(q)| = 0. \tag{3.10}$$

(b) For every $p \in S$ and every period point $q \in J$ it follows that

$$\limsup_{n\to\infty} |f^n(p) - f^n(q)| > 0.$$

In a paper with the title "Period Three Implies Chaos" (see [27]) upon which the following is essentially based they have proved the following theorem:

Theorem 3.18. *If there are three points $a, b, c \in J$ with $a < b < c$ such that $f(a) = b,\ f(b) = c$ and $f(c) = a$, then f is chaotic in the above sense.*

In connection with this theorem the following theorem is proved:

Theorem 3.19. *Under the assumptions of Theorem 3.18 there is, for every $k \in \mathbb{N}$, a point $x_k \in J$ with $f^k(x_k) = x_k$ and $f^j(x_k) \neq x_k$ for all $j = 1, \ldots, k-1$.*

For the proof of this theorem we need three lemmata.

Lemma 3.20. *Let $I, J \subseteq \mathbb{R}$ be two closed genuine intervals and $f : J \to \mathbb{R}$ a continuous function with $f(I) \supseteq J$. Then there is a closed subinterval $I_1 \subseteq I$ with $f(I_1) = J$.*

Proof. Let $J = [f(p), f(q)]$ for $p, q \in I$. If $p < q$, then there is some $\widehat{r} \in [p, q]$ with

$$\widehat{r} = \max\{r \in [p, q] \mid f(r) = f(p)\}$$

and some $\widehat{s} \in [\widehat{r}, q]$ with

$$\widehat{s} = \min\{s \in [\widehat{r}, q] \mid f(s) = f(q)\}.$$

If one puts $I_1 = [\widehat{r}, \widehat{s}]$, then $I_1 \subseteq I$ and $f(I_1) = J$. Similar arguments apply in the case $p > q$. □

Lemma 3.21. *Let $J \subseteq I\!R$ be a closed genuine interval and let $f : J \to J$ be continuous. Further let $(I_n)_{n\in I\!N_0}$ be a sequence of closed genuine intervals $I_n \subseteq J$ with $I_{n+1} \subseteq f(I_n)$ for all $n \in I\!N_0$. Then there is a sequence $(Q_n)_{n\in I\!N_0}$ of closed intervals $Q_n \subseteq J$ with*

$$Q_n \subseteq Q_{n-1} \subseteq I_0 \quad \text{and} \quad f^n(Q_n) = I_n \qquad \text{for all} \quad n \in I\!N_0$$

which implies that

$$f^n(x) \in I_n \quad \text{for all} \quad n \in I\!N_0 \qquad \text{and} \qquad x \in Q = \bigcap_{n\in I\!N_0} Q_n.$$

Proof. If we put $Q_0 = I_0$, then it follows that $f^0(Q_0) = I_0$. Now let Q_{n-1} for some $n \in \mathbb{N}$ be constructed such that

$$Q_{n-1} \subseteq I_0 \qquad \text{and} \qquad f^{n-1}(Q_{n-1}) = I_{n-1}.$$

By assumption we have $I_n \subseteq f^n(Q_{n-1})$. Lemma 3.20 with $I = Q_{n-1}$, $J = I_n$ and f^n instead of f then implies the existence of a closed interval $Q_n \subseteq Q_{n-1} \subseteq I_0$ with $f^n(Q_n) = I_n$. The existence of the sequence $(Q_n)_{n\in\mathbb{N}_0}$ with the above properties follows by the principle of induction. □

Lemma 3.22. *Let $I \subseteq J \subseteq I\!R$ be closed genuine intervals. Further let $g : J \to I\!R$ be continuous and let $I \subseteq g(I)$. Then there is a point $\widehat{x} \in I$ with $g(\widehat{x}) = \widehat{x}$.*

Proof. Let $I = [\beta_0, \beta_1]$. Then choose $\alpha_0, \alpha_1 \in I$ such that $\beta_0 = g(\alpha_0)$ and $\beta_1 = g(\alpha_1)$. Then it follows that

$$g(\alpha_0) - \alpha_0 \leq 0 \qquad \text{and} \qquad g(\alpha_1) - \alpha_1 \geq 0.$$

Since g is continuous there exists some $\widehat{x} \in I$ with $g(\widehat{x}) - \widehat{x} = 0$ which completes the proof. □

After these preparations we come to the

Proof of Theorem 3.19. The assumptions of the theorem imply

$$[b,c] \subseteq f([a,b]) \qquad \text{and} \qquad [a,c] \subseteq f([b,c]) \tag{$*$}$$

by virtue of the intermediate value theorem for continuous functions.

Since $[b,c] \subseteq [a,c]$, it follows that $[b,c] \subseteq f([b,c])$. By Lemma 3.22 we obtain the existence of some $x_1 \in [b,c]$ with $f(x_1) = x_1$. Therefore the assertion is true for $k = 1$. Now let $k > 1$ be given.

Then we define a sequence $(I_n)_{n\in\mathbb{N}_0}$ of intervals $I_n \subseteq J$ by

$$I_n = [b,c] \quad \text{for} \quad n = 0,\ldots,k-2, \qquad I_{k-1} = [a,b]$$

and

$$I_{n+k} = I_n \qquad \text{for} \quad n = 0,1,2,\ldots$$

From $(*)$ it then follows that $I_{n+1} \subseteq f(I_n)$ for all $n \in \mathbb{N}_0$.

By Lemma 3.21 there exists a closed interval $Q_k \subseteq I_0$ with $f^k(Q_k) = I_k = I_0$. By Lemma 3.22 with $g = f^k$, $I = Q_k$, $J = I_0$ there exists some $x_k \in Q_k$ with $f^k(x_k) = x_k$. Without loss of generality we can assume that $k \neq 3$.

Assumption: $f^j(x_k) = x_k$ for some $j \in \{1,\ldots,k-1\}$. Because of $f^j(Q_j) = I_j = I_0$ for $j = 0,\ldots,k-2$ it follows that

$$f^{k-1}(x_k) = f^{k-j-1} \circ f^j(x_k) = f^{k-j-1}(x_k) \in I_0 = [b,c].$$

On the other hand we have

$$f^{k-1}(x_k) \in f^{k-1}(Q_{k-1}) = I_{k-1} = [a,b].$$

This implies $f^{k-1}(x_k) = b$ and therefore $x_k = f^k(x_k) = c$. This is only possible for $k = 3$. Therefore the assumption is false for $k \neq 3$ and Theorem 3.19 is proved. □

This theorem is contained in the

Theorem of Sarkovski: *Let $f : \mathbb{R} \to \mathbb{R}$ be continuous and have a period point of period k. Then f also has, for every $l \in \mathbb{N}$ with $k \rhd l$, a period point with period l where "$\rhd$" denotes an ordering of the natural numbers by virtue of*

$$\underbrace{3 \rhd 5 \rhd 7 \rhd \ldots}_{\text{odd numbers}} \rhd 2\cdot 3 \rhd 2\cdot 5 \rhd 2\cdot 7 \rhd \ldots \rhd 2^2\cdot 3 \rhd 2^2\cdot 5 \rhd 2^2\cdot 7 \rhd \ldots$$
$$\ldots \rhd 2^3\cdot 3 \rhd 2^3\cdot 5 \rhd 2^3\cdot 7 \rhd \ldots \rhd 2^3 \rhd 2^2 \rhd 2 \rhd 1.$$

We will not give the proof of this theorem and refer to the book "Introduction to Chaotic Dynamics" by Robert L. Devaney (see [5]).

An example for Theorem 3.19 is the quadratic function

$$f(x) = \mu x(1-x), \; x \in J = [0,1] \qquad \text{for} \quad \mu \approx 3.839.$$

One confirms by calculation that f has a period three which is given approximately by

$$a = 0.149888, \quad b = 0.489172, \quad c = 0.959299.$$

Proof of Theorem 3.18. Let $\mathcal{M}$ be the set of all sequences $M = (M_n)_{n\in\mathbb{N}}$ of closed intervals with

$$M_n = [a,b] \qquad \text{or} \qquad M_n \subseteq [b,c] \quad \text{and} \quad f(M_n) \supseteq M_{n+1}. \tag{A.1}$$

If $M_n = [a,b]$, then n is the square of a natural number and

$$M_{n+1}, M_{n+2} \subseteq [b,c]. \tag{A.2}$$

If n is the square of a natural number, then this is not the case for $n+1$ and $n+2$ so that the requirement $M_{n+1}, M_{n+2} \subseteq [b,c]$ in (A.2) is redundant.

For every $M \in \mathcal{M}$ let $P(M,n)$ be the number of those $i \in \{1,\ldots,n\}$ with $M_i = [a,b]$. For every $r \in \left(\frac{3}{4},1\right)$ let $M^r = (M_n^r)_{n\in\mathbb{N}}$ be a sequence in $\mathcal{M}$ with

$$\lim_{n\to\infty} \frac{P\left(M_n^r, n^2\right)}{n} = r. \tag{A.3}$$

Let $\mathcal{M}_0 = \left\{ M^r \,\middle|\, r \in \left(\frac{3}{4},1\right) \right\}$. Then $\mathcal{M}_0$ is uncountable, since $M^{r_1} \neq M^{r_2}$ for $r_1 \neq r_2$. By Lemma 3.21 there exists for every $M^r \in \mathcal{M}_0$ a point $x_r \in [a,c]$ with

$$f^n(x_r) \in M_n^r \qquad \text{for all} \quad n \in \mathbb{N}.$$

Put

$$S = \left\{ x_r \in [a,c] \,\middle|\, r \in \left(\frac{3}{4},1\right) \right\}.$$

Then S is uncountable. For every $x \in S$ let $P(x,n)$ be the number of those $i \in \{1,\ldots,n\}$ for which $f^i(x) \in [a,b]$. Certainly we never have $f^k(x_r) = b$ for $x_r \in S$, since otherwise f^k would have a period 3 which because of (A.2) is impossible.

This implies

$$P(x_r,n) = P(M^r,n) \qquad \text{for all} \quad n \in \mathbb{N}$$

and therefore

$$\varrho(x_r) = \lim_{n\to\infty} \frac{P\left(x_r,n^2\right)}{n} = r \qquad \text{for all} \quad r \in \left(\frac{3}{4},1\right).$$

Assertion: For $p,q \in S$ with $p \neq q$ there are infinitely many $n \in \mathbb{N}$ such that

$$f^n(p) \in [a,b] \quad \text{and} \quad f^n(q) \in [b,c] \quad \text{or vice versa.} \tag{A.4}$$

We can assume that $\varrho(p) > \varrho(q)$. Then it follows that

$$\lim_{n\to\infty} \left(P(p,n) - P(q,n)\right) = \infty.$$

This implies that there are infinitely many $n \in \mathbb{N}$ with (A.4).

Since $f^2(b) = a$ and f^2 is continuous, there exists some $\delta > 0$ such that

$$f^2(x) < \frac{a+b}{2} \qquad \text{for all} \quad x \in [b-\delta, b] \subseteq [a,b].$$

If $p \in S$ and $f^n(p) \in [a,b]$, then (A.2) implies

$$f^{n+1}(p) \in [b,c] \qquad \text{and} \qquad f^{n+2}(p) \in [b,c]$$

and hence $f^n(p) < b - \delta$.

If $f^n(q) \in [b,c]$, then $f^n(q) \geq b$ and therefore

$$|f^n(p) - f^n(q)| > \delta.$$

By Assertion (A.4) we conclude that for every pair $p, q \in S$ with $p \neq q$ it follows that

$$\limsup_{n \to \infty} |f^n(p) - f^n(q)| > \delta.$$

This shows that (3.9) is true.

In order to prove (b) we at first show that S does not contain period points. Let $p \in J$ be a period point of the period $k \in \mathbb{N}$. Then we distinguish the following two cases:

(α) $\{f(p), \ldots, f^k(p)\} \cap [a,b] = \emptyset$.
Then it follows that $P(p, n^2) = 0$ for all $n \in \mathbb{N}$ and hence $\varrho(p) = 0$. This shows that p cannot belong to S, since for every $x_r \in S$ we have $\varrho(x_r) \in \left(\frac{3}{4}, 1\right)$.

(β) $\{f(p), \ldots, f^k(p)\} \cap [a,b] \neq \emptyset$.
Let $\widehat{k}$ be the number of those $j \in \{1, \ldots, k\}$ with $f^j(p) \in [a,b]$. If we choose in particular $n = n_l = l \cdot k$ for $l \in \mathbb{N}$, then it follows that

$$\frac{P(p, n^2)}{n} = \frac{P(p, n_l^2)}{n_l} = \frac{l^2\, k}{lk} \widehat{k} = l\widehat{k} \to \infty$$

for $l \to \infty$. Therefore it is impossible that

$$\lim_{n \to \infty} \frac{P(p, n^2)}{n} \in \left(\frac{3}{4}, 1\right)$$

which shows that $p \notin S$.

Result: S does not contain period points.

Now let $p \in J$ be a period point and let $q \in S$ be given. We at first assume the case (α), i.e.,

$$f^j(p) \notin [a,b] \qquad \text{for all} \quad j \in \mathbb{N}.$$

Now let

$$f^j(p) < a \qquad \text{for some} \quad j \in \{1, \dots, k\}.$$

Then we put

$$\delta = \min\left\{|a - f^j(p)| \mid f^j(p) < a\right\}.$$

Since $f^n(q) \in [a, c]$ for all $n \in \mathbb{N}$, it follows that

$$|f^n(p) - f^n(q)| > \delta \qquad \text{for infinitely many} \quad n.$$

This implies

$$\limsup_{n\to\infty} |f^n(p) - f^n(q)| > \delta.$$

Let

$$f^j(p) > b \qquad \text{for all} \quad j \in \{1, \dots, k\}.$$

Then it follows that $f^n(p) > b$ for all $n \in \mathbb{N}$.

Let us put

$$\delta = \min\left\{f^j(p) - b \mid j = 1, \dots, k\right\}.$$

Since $f^n(q) \in [a, b]$ for infinitely many n, it follows that

$$|f^n(p) - f^n(q)| > \delta \qquad \text{for infinitely many} \quad n.$$

This implies

$$\limsup_{n\to\infty} |f^n(p) - f^n(q)| > \delta.$$

In case (β) we infer from $\varrho(p) = \infty$ that

$$\lim_{n\to\infty} (P(p, n) - P(q, n)) = \infty.$$

This implies that there are infinitely many $n \in \mathbb{N}$ with (A.4).

The proof can now be continued as the proof of (3.9). For the proof of (3.10) we at first observe that because of $f(b) = c$ and $f(c) = a$ we can construct intervals $[b^n, c^n]$, $n = 0, 1, 2, \dots$ with the following properties (see Lemma 3.20):

(α) $[b, c] = \left[b^0, c^0\right] \supseteq \left[b^1, c^1\right] \supseteq \dots \supseteq [b^n, c^n] \supseteq \dots,$

(β) $f(x) \in (b^n, c^n)$ for all $x \in \left(b^{n+1}, c^{n+1}\right),$

(γ) $f\left(b^{n+1}\right) = c^n,\ f\left(c^{n+1}\right) = b^n.$

Let $A = \bigcap_{n=0}^{\infty} [b^n, c^n]$, $b^* = \inf A$ and $c^* = \sup A$. Then it follows that $f(b^*) = c^*$ and $f(c^*) = b^*$ as a consequence of (γ).

In order to prove (3.10) we have to make a careful choice of the sequences $M^r \in \mathcal{M}$. In addition to the above requirements we assume the following:

If $M_k = [a,b]$ for some $k = n^2$ and $k = (n+1)^2$, then we choose

$$M_{n^2+(2j-1)} = \left[b^{2n-(2j-1)}, b^*\right] \qquad \text{and} \qquad M_{n^2+2j} = \left[c^*, c^{2n-2j}\right]$$

for $j = 1, \ldots, n$. For the remaining $k \in \mathbb{N}$ which are no squares we choose $M_k = [b,c]$.

One can check that these additional requirements are consistent with (A.1) and (A.2) and that for every $r \in (3/4, 1)$ one can find a sequence $M^r \in \mathcal{M}$ which satisfies (A.3).

From $\varrho(r) = r$ for all $r \in \left(\frac{3}{4}, 1\right)$ it follows for $r, r^* \in \left(\frac{3}{4}, 1\right)$ the existence of infinitely many $n \in \mathbb{N}$ with $M_k^r = M_k^{r^*} = [a,b]$ for $k = n^2$ and $k = (n+1)^2$.

Now let x_r and $x_{r^*} \in S$ be given. Because of $b^n \to b^*$ and $c^n \to c^*$ for $n \to \infty$, for every $\varepsilon > 0$, there is some $N \in \mathbb{N}$ with

$$|b^n - b^*| < \frac{\varepsilon}{2} \quad \text{and} \quad |c^n - c^*| < \frac{\varepsilon}{2} \qquad \text{for all} \quad n > N.$$

For every $n \in \mathbb{N}$ with $n > N$ and $M_k^r = M_k^{r^*} = [a,b]$ for $k = n^2$ and $k = (n+1)^2$ it therefore follows that

$$f^{n^2+1}(x_r) \in M_k^r = \left[b^{2n-1}, b^*\right] \qquad \text{and} \qquad f^{n^2+1}(x_{r^*}) \in M_k^{r^*} = \left[b^{2n-1}, b^*\right]$$

for $k = n^2 + 1$ which implies $\left|f^{n^2+1}(x_r) - f^{n^2+1}(x_{r^*})\right| < \varepsilon$. From this it follows that

$$\liminf_{n\to\infty} |f^n(x_r) - f^n(x_{r^*})| = 0$$

which concludes the proof of Theorem 3.18. □

With the aid of the Sarkovski Theorem it is possible to derive from Theorem 3.18 the following

Theorem 3.23. *Let $J \subseteq \mathbb{R}$ be a finite closed interval and let $f : J \to J$ be a continuous function with a period point of period k with $k \neq 2^i$, $i \in \mathbb{N}$. Then f is chaotic in the sense of Li and Yorke.*

Proof. Because of $k \neq 2^i$, $i \in \mathbb{N}$, k can be represented in the form $k = mp$ with $m \in \mathbb{N}$ and $p > 2$ being a prime number. This implies that f^m is p−periodic. By the Theorem of Sarkovski it follows that f^m has a period point of period $2 \cdot 3$ and hence f^{2m} has a period point of period 3. By Theorem 3.18 (which is also true without the assumption $a < b < c$) f^{2m} is chaotic in the sense of Li and Yorke. This implies that f is also chaotic in the sense of Li and Yorke. □

With the aid of Theorem 3.23 one can also gain a connection between disorder-chaos and chaos in the sense of Li and Yorke. For that purpose we need the statement that a continuous mapping f of a finite closed interval J into itself with positive topological entropy has a period point of period k with $k \neq 2^i$, $i \in \mathbb{N}$. This was proved by M. Misiurewicz in [29] and [30], however, with a formally different definition of entropy. An immediate consequence of this statement and Theorem 3.23 is the statement that disorder chaos (for one-dimensional mappings) implies chaos in the sense of Li and Yorke.

3.5 Strange (or Chaotic) Attractors

This section is essentially based on representations in the book [5] of Devaney. In Section 1.3.3 we have introduced the concept of an attractor which will be slightly modified here. We assume a continuous mapping $f : D \to \mathbb{R}^n$, $D \subseteq \mathbb{R}^n$ to be given.

Definition. *A subset $A \subseteq D$ is called an attractor of f, if there exists a subset $V \subseteq D$ with $f(V) \subseteq V$ and $\overline{A} \subseteq V$ such that*

$$\overline{A} = \bigcap_{n=0}^{\infty} f^n(V)$$

where

$$f^n(V) = \{f^n(x) \mid x \in V\} \qquad \text{and} \quad \overline{A} = \text{ closure of } A.$$

Lemma 3.24. *If $A \subseteq D$ is an attractor of f, then $\overline{A}$ is invariant, i.e., $f\left(\overline{A}\right) = \overline{A}$.*

Proof. On the one hand we have

$$f\left(\overline{A}\right) = f\left(\bigcap_{n=0}^{\infty} f^n(V)\right) = \bigcap_{n=1}^{\infty} f^n(V) \supseteq \overline{A},$$

on the other hand it follows that

$$f\left(\overline{A}\right) = \bigcap_{n=1}^{\infty} f^{n-1}(f(V)) \subseteq \bigcap_{n=1}^{\infty} f^{n-1}(V) = \bigcap_{n=0}^{\infty} f^n(V) = \overline{A}.$$

hence $f\left(\overline{A}\right) = \overline{A}$. □

Lemma 3.25. *Let $A \subseteq D$ be an attractor of f such that there is some $n_0 \in \mathbb{N}$ such that*

$$f^n(V) \quad \text{is closed for all} \quad n \geq n_0.$$

Then it follows that

$$L_f(x) \subseteq A \qquad \text{for all} \quad x \in V$$

where $L_f(x)$ is the limit set of the orbit $\gamma_f(x)$ (1.42) given by (1.43).

Proof. By Proposition 1.16 it follows that

$$L_f(x) = \left\{ y \in \mathbb{R}^n \;\middle|\; \exists \text{ monotone sequence } n_i \to \infty \text{ with } y = \lim_{n_i \to \infty} f^{n_i}(x) \right\}.$$

Now let $x \in V$ and $y \in L_f(x)$ be given. Let $n_i \to \infty$ be a monotone sequence with $y = \lim_{i\to\infty} f^{n_i}(x)$. For every $i \in \mathbb{N}$ we have

$$x_{n_i} = f^{n_i}(x) \in f^{n_i}(V)$$

and it follows that $x_{n_i} \to y$.

By assumption $f^{n_i}(V)$ is closed for sufficiently large i. Because of

$$f^{n+1}(V) \subseteq f^n(V) \qquad \text{for all} \quad n \in \mathbb{N}_0$$

it therefore follows that

$$A = \overline{A} = \bigcap_{i\in\mathbb{N}} f^{n_i}(V).$$

Further it follows that $y \in f^{n_i}(V)$ for all sufficiently large i and therefore

$$y \in \bigcap_{i\in\mathbb{N}} f^{n_i}(V) = \overline{A} = A.$$

□

Definition. *An attractor $A \subseteq D$ of f is called strange (or chaotic), if $f : A \to A$ is topologically conjugated to the shift-mapping $\sigma : \Sigma \to \Sigma$ of the metric space Σ of the $0-1-$sequences (see Section 3.1) into itself.*

Since the shift-mapping is chaotic in the sense of Devaney (see Theorem 3.6, the Corollary 3.8 leads to the following

Theorem 3.26. *If $A \subseteq D$ is a strange attractor of f, then $f : \overline{A} \to \overline{A}$ is chaotic in the sense of Devaney.*

Examples:

(a) Smale's Horse-Shoe-Mapping: A classical example of a chaos generating mapping in the plane is Smale's horse-shoe-mapping f_H which can be illustrated by the following picture:

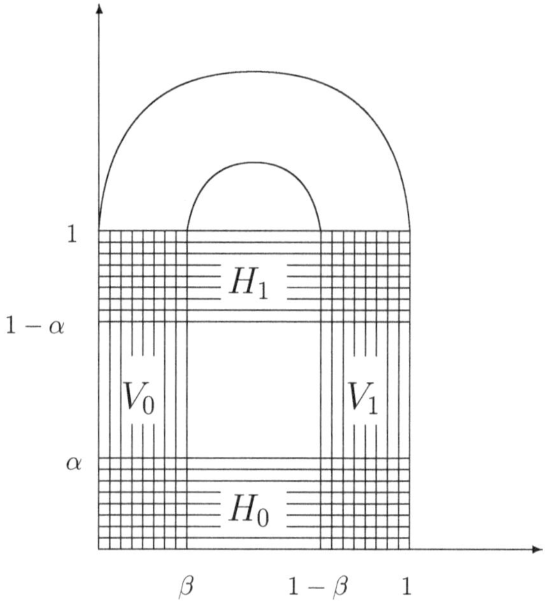

The mapping f_H maps the square $Q = [0,1]\times[0,1]$ onto the horse-shoe in such a way that the horizontal strips $H_0 = [0,1]\times[0,\alpha]$ and $H_1 = [0,1]\times[1-\alpha]$ are mapped onto the vertical strips $V_0 = [0,\beta]\times[0,1]$ and $V_1 = [1-\beta,1]\times[0,1]$ with $0 < \beta \le \alpha < 1/2$ by virtue of

$$\left\{\begin{matrix} f_H(x,y) = \left(\beta x, \dfrac{1}{\alpha}y\right) \\ \text{for} \quad 0 \le x \le 1,\ 0 \le y \le \alpha \end{matrix}\right\} \quad \text{and} \quad \left\{\begin{matrix} f_H(x,y) = \left(1-\beta x, \dfrac{1}{\alpha} - \dfrac{1}{\alpha}y\right) \\ \text{for} \quad 0 \le x \le 1,\ 1-\alpha \le y \le 1 \end{matrix}\right\},$$

respectively. The horizontal strip between H_0 and H_1 is mapped in a suitable way onto the circular strip which together with V_0 and V_1 forms the horse-shoe.

Obviously we have

$$\begin{aligned} f_H^{-1}(Q) &= Q \setminus [0,1]\times[\alpha,1-\alpha] = H_0 \cup H_1, \\ f_H^{-2}(Q) &= f_H^{-1}(H_0 \cup H_1) = \left(H_0^1 \cup H_0^2\right) \cup \left(H_1^1 \cup H_1^2\right) \end{aligned}$$

and in general for every $n \in \mathbb{N}$ it follows that

$$f_H^{-n}(Q) = \left(\bigcup_{i=1}^{2^{n-1}} H_0^i \cup \bigcup_{i=1}^{2^{n-1}} H_1^i \right)$$

where every H_0^i and H_1^i is a horizontal strip in H_0 and H_1, respectively, which is obtained by removing a horizontal strip of relative length $1-2\alpha$ from H_0^{i-1} and H_1^{i-1}, respectively.

If one defines

$$V = \bigcap_{n=1}^{\infty} f_H^{-n}(Q),$$

then V is representable in the form $V = [0,1] \times C_\alpha$ where C_α is a Cantor set which is obtained from $[0,1]$ by permanently removing open intervals of relative length $1 - 2\alpha$.

All sets $f_H^{-n}(Q)$ are closed subsets of Q and therefore even compact. This implies that V is closed and therefore also compact. Further we have

$$f_H(V) = \bigcap_{n=1}^{\infty} f_H\left(f_H^{-n}(Q)\right) = \bigcap_{n=1}^{\infty} f_H^{1-n}(Q) = \bigcap_{m=0}^{\infty} f_H^{-m}(Q) \subseteq V.$$

If one defines $A = \bigcap_{n=0}^{\infty} f_H^{n}(V)$, then it follows that $A = \overline{A}$, since all sets $f_H^n(V)$ are compact and therefore closed. Therefore $A \subseteq Q$ is an attractor of the mapping $f_H : Q \to \mathbb{R}^2$. Lemma 3.25 implies

$$L_{f_H}(x) \subseteq A \qquad \text{for all} \quad x \in V$$

where $L_{f_H}(x)$ is the limit set of the orbit $\gamma_{f_H}(x)$ (1.42) given by (1.43) with $f_H : V \to V$.

Now let Σ_2 be the metric space of twofold infinite $0-1$–sequences, i.e.,

$$\Sigma_2 = \left\{ s = (s_j)_{j \in \mathbb{Z}} \mid s_j = 0 \ \text{ or } \ 1 \right\}$$

with metric

$$d(s,t) = \sum_{j \in \mathbb{Z}} \frac{1}{2^{|j|}} |s_j - t_j|, \qquad s, t \in \Sigma_2.$$

In Section 3.2 we have shown that the shift-mapping $\sigma : \Sigma_2 \to \Sigma_2$ given by

$$\sigma(s)_j = s_{j+1} \qquad \text{for all} \quad j \in \mathbb{Z} \ \text{ and all } \ s \in \Sigma_2$$

(which is a homeomorphism) is chaotic in the sense of Devaney.

Now let us define a mapping $S : A \to \Sigma_2$ by $s = S(x),\ x \in A$, where for every $j \in \mathbb{Z}$ we define

$$s_j = \begin{cases} 0, & \text{if} \quad f_H^j(x) \in V_0, \\ 1, & \text{if} \quad f_H^j(x) \in V_1. \end{cases}$$

From

$$S(x)_{j+1} = \left\{ \begin{matrix} 0, & \text{if} \quad f_H^{j+1}(x) \in V_0, \\ 1, & \text{if} \quad f_H^{j+1}(x) \in V_1, \end{matrix} \right\}$$

$$= \left\{ \begin{matrix} 0, & \text{if} \quad f_H^j\left(f_H(x)\right) \in V_0, \\ 1, & \text{if} \quad f_H^j\left(f_H(x)\right) \in V_1 \end{matrix} \right\} = S\left(f_H(x)\right)_j$$

it follows that

$$\sigma(S(x)) = S\left(f_H(x)\right) \qquad \text{for all} \quad x \in A.$$

Lemma 3.27. *$S : A \to \Sigma_2$ is a homeomorphism.*

Proof. At first we observe that A can be represented in the form $A = V \cap (C_\beta \times [0,1])$ where C_β is a Cantor set which is obtained from $[0,1]$ by permanently removing open intervals of relative length $1 - 2\beta$. In particular we have $A \subseteq V_0 \cup V_1$ so that because of $f_H(A) = A$ the mapping $S : A \to \Sigma_2$ is well defined.

Next we prove the injectivity of $S : A \to \Sigma_2$. Let $S(x) = S(y)$ for $x, y \in A$. Then it follows that for every $j \in \mathbb{Z}$ that

$$\left(f_H^j(x) \in V_0 \ \text{ and } \ f_H^j(y) \in V_0\right) \quad \text{or} \quad \left(f_H^j(x) \in V_1 \ \text{ and } \ f_H^j(y) \in V_1\right)$$

Because of the linearity of $f_H^j : A \to A$ for all $j \in \mathbb{Z}$ and the convexity of V_0 and V_1 it follows for every $j \in \mathbb{Z}$ and $\lambda \in [0,1]$ that

$$f^j(\lambda x + (1-\lambda)y) \in V_0 \quad \text{or} \quad f^j(\lambda x + (1-\lambda)y) \in V_1.$$

However, this is only possible, if $\lambda = 0$ or 1 or $x = y$, since $[x, y] \nsubseteq A$, if $x \neq y$. This follows from the fact that $A = V \cap (C_\beta \times [0,1])$ and that C_β does not contain intervals.

In order to prove surjectivity of $S : A \to \Sigma_2$ we consider some $s = (\ldots, s_{-n}, \ldots, s_0, \ldots, s_n, \ldots)$ in Σ_2. For every $n \in \mathbb{N}_0$ we define

$$I_n(s) = \bigcap_{j=-n}^{n} f_H^{-j}\left(V_{s_j}\right).$$

Every I_n, $n \in \mathbb{N}_0$, is non-empty and closed. Further we have $I_{n+1} \subseteq I_n$ for all $n \in \mathbb{N}_0$ which implies that

$$I_\infty(s) = \bigcap_{n \in \mathbb{N}_0} I_n(s) \subseteq A$$

is non-empty, since all I_n are contained in the compact set Q. Let $x \in I_\infty(s)$.

Then it follows that

$$f_H^j(x) \in V_{s_j} \text{ for all } j \in \mathbb{N}_0 \quad \Longleftrightarrow \quad s = S(x).$$

Because of the injectivity of $S : A \to \Sigma_2$ it follows that $I_\infty(s) = \{x\}$.

Finally we show the continuity of S. For that purpose we choose some $x \in A$ arbitrarily and put $(\ldots, s_{-n}, \ldots, s_0, \ldots, s_n, \ldots) = S(x)$. Let $\varepsilon > 0$ be given. Then we choose $n \in \mathbb{N}_0$ such that $\frac{1}{2^{n-1}} \leq \varepsilon$. Now we consider the set of all $I_n(t)$ for all 2^{2n+1} combinations of $t_{-n}, t_{-n+1}, \ldots, t_{n-1}, t_n$. These are all different and $I_n(s)$ belongs to this set. Now we choose $\delta > 0$ such that

$$\|x - y\|_2 < \delta \text{ and } y \in A \quad \Longrightarrow \quad y \in I_n(s).$$

Then it follows that

$$S(x)_i = S(y)_i \qquad \text{for} \quad i = -n, \ldots, n.$$

This implies $d(S(x), S(y)) \leq \frac{1}{2^{n-1}} \leq \varepsilon$ (see Section 3.2) which shows the continuity of S. □

This implies the continuity of the inverse mapping S^{-1} and shows that $\sigma : \Sigma_2 \to \Sigma_2$ and $f_H : A \to A$ are topologically conjugated. The Corollary 3.8 then implies that f_H is chaotic in the sense of Devaney. If we replace in the definition of a strange attractor Σ by Σ_2 which is not an essential change, then it also follows that

$$A = \bigcap_{n=0}^{\infty} f_H^n(V) \qquad \text{with} \qquad V = \bigcap_{n=1}^{\infty} f_H^{-n}(Q)$$

is a strange attractor.

(b) The Solenoid: We consider the unit circle

$$S^1 = \left\{ \mathrm{e}^{it} \mid t \in [0, 2\pi] \right\}, \qquad i = \sqrt{-1},$$

and the unit circle disk

$$B^2 = \left\{ (x, y) \in \mathbb{R}^2 \mid x^2 + y^2 \leq 1 \right\}.$$

With these two we define a torus $D = S^1 \times B^2$. On D we define a continuous mapping $f : D \to D$ by

$$f\left(\mathrm{e}^{it}, p\right) = \left(\mathrm{e}^{2it}, \frac{1}{10} p + \frac{1}{2} \mathrm{e}^{2it} \right) \qquad \text{for} \quad \mathrm{e}^{it} \in S^1 \text{ and } p \in B^2.$$

Geometrically f can be described as follows: Let $\mathrm{e}^{it^*} \in S^1$ be given. Then we denote by $B(t^*)$ the disk

$$B(t^*) = \left\{ \left(\mathrm{e}^{it^*}, p \right) \mid p \in B^2 \right\}.$$

With this notation we have $f(B(t)) \subseteq B(2t)$. The exact image $f(B(t))$ is a disk with the center

$$\left(\mathrm{e}^{2it}, \frac{1}{2}\mathrm{e}^{2it} \right) \quad \text{and the radius } \frac{1}{10}.$$

On the whole we obtain the picture

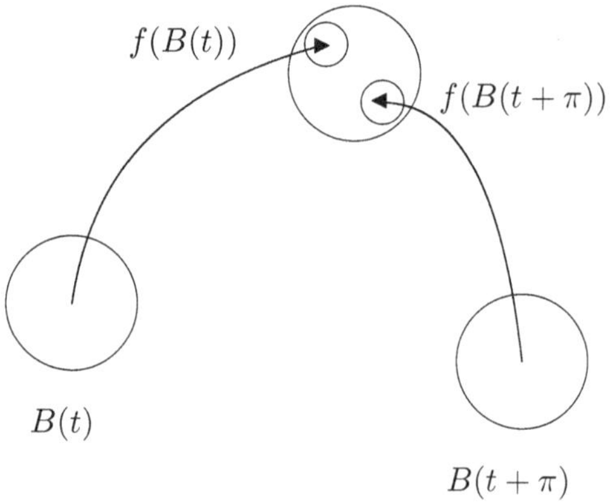

The image of D consists of two tori in the interior of D. The image $f^2(D)$ of $f(D)$ consists of four tori, two of them, respectively, contained in the two tori of which $f(D)$ consists. The radii of this these four tori are equal to $\frac{1}{100}$. In general we obtain

$$f^n(D) \subseteq f^{n-1}(D) \subseteq \ldots \subseteq f(D) \subseteq D \qquad \text{for every} \quad n \in \mathbb{N}$$

and every set $f^k(D)$ is closed. This implies that

$$A = \bigcap_{n \in \mathbb{N}_0} f^n(D)$$

is non-empty and is a closed attractor. It is called a solenoid. For this attractor one can prove the following

Theorem 3.28.
(a) The set Per(f) *of the period points of* $f : A \to A$ *is dense in* A.
(b) f *is topologically transitive on a suitable compact subset of* A.

Proof.
(a) Let $x = (\mathrm{e}^{it}, p_0) \in A$ be given. Further let $U \subseteq D$ be any open neighborhood of x. Then there is some $\delta > 0$ and some $n \in \mathbb{N}_0$ such that the tube

$$C = \left\{\left(\mathrm{e}^{it}, p\right) \in f^n(A) \mid |t - t_0| < \delta\right\}$$

is contained in U. Since the period points of the mapping $\mathrm{e}^{it} \to \mathrm{e}^{2it}$, $t \in [0, 2\pi]$, are of the form $\mathrm{e}^{it_{mk}}$ with $m \in \mathbb{N}$ and $t_{mk} = \dfrac{2k\pi}{2^m - 1}$ for $k = 0, \ldots, 2^m - 1$, they are lying dense in S^1. Therefore there is some $t^* \in \{t \in [0, 2\pi] \mid |t - t_0| < \delta\}$ with $f^m(B(t^*)) \subseteq B(t^*)$. We can assume $m \in \mathbb{N}$ to be so large that $f^m(B(t^*) \cap C) \subseteq B(t^*) \cap C$. Since $f^m = f^m(\mathrm{e}^{it}, p)$ is contractive with respect to p, there exists a point $\left(\mathrm{e}^{it^*}, p^*\right) \in B(t^*) \cap C$ with $f^m\left(\mathrm{e}^{it^*}, p^*\right) = \left(\mathrm{e}^{it^*}, p^*\right) \in A$ which completes the proof.
We will not give a direct proof of (b). The statement (b) will be the result of the following considerations:
Let $g : S^1 \to S^1$ be defined by

$$g\left(\mathrm{e}^{it}\right) = \mathrm{e}^{2it}, \qquad t \in [0, 2\pi].$$

Further we define the sequence space

$$\Sigma = \left\{\Theta_t = \left(\mathrm{e}^{it}, \mathrm{e}^{i\frac{t}{2}}, \ldots, \mathrm{e}^{i\frac{t}{2^n}}, \ldots\right) \middle| \, t \in [0, 2\pi]\right\}.$$

If we define a metric in Σ by

$$d(\Theta_t, \Theta_s) = \sum_{j=0}^{\infty} \frac{1}{2^j} \left|\mathrm{e}^{i\frac{t}{2^j}} - \mathrm{e}^{i\frac{s}{2^j}}\right|, \qquad \Theta_t, \Theta_s \in \Sigma,$$

then Σ becomes a metric space. On Σ there exists a natural mapping

$$\sigma\left(\mathrm{e}^{it}, \mathrm{e}^{i\frac{t}{2}}, \mathrm{e}^{i\frac{t}{4}}, \ldots\right) = \left(g\left(\mathrm{e}^{i\frac{t}{2}}\right), \mathrm{e}^{it}, \mathrm{e}^{i\frac{t}{2}}, \ldots\right), \qquad t \in [0, 2\pi],$$

which maps Σ continuously into itself and whose inverse mapping is given by

$$\sigma^{-1}\left(\mathrm{e}^{it}, \mathrm{e}^{i\frac{t}{2}}, \mathrm{e}^{i\frac{t}{4}}, \ldots\right) = \left(\mathrm{e}^{i\frac{t}{2}}, \mathrm{e}^{i\frac{t}{4}}, \mathrm{e}^{i\frac{t}{8}}, \ldots\right), \qquad t \in [0, 2\pi],$$

and is also continuous.

Therefore $\sigma : \Sigma \to \Sigma$ is a homeomorphism.

If we define a continuous mapping $\varphi : S^1 \to \Sigma$ by

$$\varphi\left(\mathrm{e}^{it}\right) = \left(\mathrm{e}^{it}, \mathrm{e}^{i\frac{t}{2}}, \mathrm{e}^{i\frac{t}{4}}, \ldots\right), \qquad t \in [0, 2\pi]$$

then it follows that φ is surjective. Further it follows that

$$\sigma \circ \varphi\left(\mathrm{e}^{it}\right) = \sigma\left(\mathrm{e}^{it}, \mathrm{e}^{i\frac{t}{2}}, \mathrm{e}^{i\frac{t}{4}}, \ldots\right) = \left(\mathrm{e}^{2it}, \mathrm{e}^{it}, \mathrm{e}^{i\frac{t}{2}}, \ldots\right) \quad \text{and}$$

$$\varphi \circ g\left(\mathrm{e}^{it}\right) = \varphi\left(\mathrm{e}^{2it}\right) = \left(\mathrm{e}^{2it}, \mathrm{e}^{it}, \mathrm{e}^{i\frac{t}{2}}, \ldots\right) = \sigma \circ g\left(\mathrm{e}^{it}\right) \quad \text{for all } t \in [0, 2\pi],$$

hence $\varphi \circ g = \sigma \circ \varphi$. Since g is chaotic in the sense of Devaney (see Section 3.2), it follows from Theorem 3.7 that $\sigma : \Sigma \to \Sigma$ is also chaotic in the sense of Devaney.

Now let $\pi : D \to S^1$ be the natural projection, i.e.,

$$\pi\left(\mathrm{e}^{it}, p\right) = \mathrm{e}^{it} \qquad \text{for all} \quad \left(\mathrm{e}^{it}, p\right) \in D.$$

The mapping $f : D \to D$ is a homeomorphism of D on $f(D)$ which implies, because of $f(A) = A$, that $f : A \to A$ is a homeomorphism of A on A. Therefore the mapping $S : A \to \Sigma$ which is defined by

$$S(x) = \left(\pi(x), \pi\left(f^{-1}(x)\right), \pi\left(f^{-2}(x)\right), \ldots\right), \qquad x \in A,$$

is well defined and it follows that

$$S(f(x)) = \left(\pi(f(x)), \pi(x), \pi\left(f^{-1}(x)\right), \ldots\right) = \sigma(S(x)) \qquad \text{for all} \quad x \in A,$$

hence

$$S \circ f = \sigma \circ S.$$

In order to further investigate the mapping S we at first analyze the set A. By definition it can be represented in the form

$$A = \bigcap_{k=0}^{\infty} \bigcup_{j=1}^{2^k} T_j(k)$$

where $T_j(k)$, for every $k \in \mathbb{N}_0$ and $j \in \left\{1, \ldots, 2^k\right\}$ is a torus in the interior of D with cross section radius $\frac{1}{10^k}$. Further we have

$$f\left(T_j(k)\right) \subseteq T_j(k) \qquad \text{for every} \quad k \in \mathbb{N}_0 \text{ and } j = 1, \ldots, 2^k$$

and for every $k \in \mathbb{N}_0$ and $j \in \left\{1, \ldots, 2^k\right\}$ there exists some

$$i = i(j,k) \in \left\{1, \ldots, 2^{k+1}\right\} \qquad \text{with} \quad T_i(k+1) \subseteq T_j(k).$$

Let $J = (j_k)_{k \in \mathbb{N}_0}$ be a sequence with $j_k \in \{1, \ldots, 2^k\}$ and

$$A_J = \bigcap_{k=0}^{\infty} T_{j_k}(k).$$

This set is a (non-empty) compact subset of A and it follows that

$$f(A_J) = \bigcap_{k=0}^{\infty} f(T_{j_k}(k)) \subseteq \bigcap_{k=0}^{\infty} T_{j_k}(k) = A_J.$$

Further, for every $t \in [0, 2\pi]$ the set $A_J \cap B(t)$ consists of one point which implies that the mapping $S : A_J \to \Sigma$ is bi-unique, continuous and surjective. Because of the compactness of A_J the mapping $S : A_J \to \Sigma$ is a homeomorphism which implies that the mappings $f : A_J \to A_J$ and $\sigma : \Sigma \to \Sigma$ are topologically conjugated. From the Corollary 3.8 we infer the

Theorem 3.29. *The mapping $f : A_J \to A_J$ is chaotic in the sense of Devaney.*

This implies statement (b) of Theorem 3.28 with the subset A_J of A. □

3.6 Bibliographical Remarks

The abstract definition of a dynamical system in Section 1.1 goes back to G. D. Birkhoff (see [3]). For metric spaces X it can be found in the essay "What is a Dynamical System?" by G. R. Sell in [8] and for uniform spaces X in [33].

It is also shown there how a non-autonomous system of differential equations

$$\dot{x} = f(x,t) \qquad \text{with} \quad f \in \mathcal{C}\left(W \times \mathbb{R}, \mathbb{R}^n\right),$$

$W \subseteq \mathbb{R}^n$ open, can be assigned a dynamical system.

The Section 1.2.1 about elementary properties of time continuous dynamical systems are also taken from the above mentioned essay of G. R. Sell. This also gives a short account of the Poincaré-Bendixon theory for systems in the plane which we have extensively represented in Section 1.2.2.

The Section 1.2.3 about stability of time-continuous dynamical systems by applying the direct method of Lyapunov can also be found in [14].

The application to general predator-prey models in Section 1.2.4 is based on the two papers [18] and [20].

Section 1.3 is taken from [22] with slight modifications and corrections. The localization of limit sets with the aid of Lyapunov functions in Section 1.3.2 has been first represented by J. T. La Salle in [25] and can also be found in [14] and [22].

The stability results given in Section 1.3.3 generalize a standard result on the stability of fixed points to be found in [22], [23] (see Theorem 1.3).

Theorem 1.5 in Section 1.3.4 coincides with Satz 5.4 in [23], if we choose as normed linear space the n–space $\mathbb{R}^n$ equipped with any norm.

The results on linear systems following the Corollary of Theorem 1.6 in Section 1.3.5 and also those for non-autonomous linear systems in Section 1.3.10 are based on Section 4.2 in [23].

The stability results in Section 1.3.9 generalize those of Section 1.3.3 and are taken from [16]. Lyapunov's method has also been applied to non-autonomous systems in [1] and [24] in order to investigate stability of fixed points. Instead of one Lyapunov function a sequence of such is used.

Following the example of G. R. Sell in [8] and [33] the controlled system (2.1) can also be assigned a dynamical system. For that purpose we assume that the functions $f_i : \mathbb{R}^{n+m} \to \mathbb{R}^n$ for $i = 1, \ldots, n$ are continuous and bounded.

Every function $u \in \mathcal{C}\left(\mathbb{R}, \mathbb{R}^m\right)$ we now assign a function $f^u \in \mathcal{C}_b\left(\mathbb{R}^{n+1}, \mathbb{R}^n\right)$ =vector space of continuous and bounded functions $g : \mathbb{R}^n \times \mathbb{R} \to \mathbb{R}^n$ by defining

$$f^u(x,t) = f(x, u(t)), \qquad (x,t) \in \mathbb{R}^{n+1}$$

and put

$$X = \{f^u \mid u \in \mathcal{C}(\mathbb{R}, \mathbb{R}^m)\}.$$

Then X is a metric space with metric

$$d(f^{u_1}, f^{u_2}) = \sup_{(x,t)\in\mathbb{R}^{n+1}} \|f^{u_1}(x,t) - f^{u_2}(x,t)\|_2, \qquad f^{u_1}, f^{u_2} \in X,$$

where $||\cdot||_2$ denotes the Euclidian norm in $\mathbb{R}^n$. The system (2.1) can then be written in the form

$$\dot{x}(t) = f^u(x(t), t), \qquad t \in \mathbb{R}. \tag{2.1'}$$

By the assumption made in Section 2.1.1 there exists, for every $u \in \mathcal{C}(\mathbb{R}, \mathbb{R}^m)$ and every point $x_0 \in \mathbb{R}^n$, a solution $x = \varphi_u(x_0, t) \in \mathcal{C}^1(\mathbb{R}, \mathbb{R}^n)$ of (2.1') with $\varphi_u(x_0, 0) = x_0$. With the aid of this solution we now define a mapping $\pi : \mathbb{R}^n \times X \times \mathbb{R} \to \mathbb{R}^n \times X$ by

$$\pi(x, f^u, \tau) = (\varphi_u(x, \tau), f^u_\tau) \qquad x \in \mathbb{R}^n,\ f^u \in X,\ \tau \in \mathbb{R},$$

where

$$f^u_\tau(x,t) = f^u(x, t+\tau) \qquad \text{for all} \quad (x,t) \in \mathbb{R}^{n+1}.$$

Obviously we obtain

$$\pi(x, f^u, 0) = (x, f^u) \qquad \text{for all} \quad (x, f^u) \in \mathbb{R}^n \times X,$$

i.e., the identity property holds true.

If one puts

$$\varphi(\tau) = \varphi_u(x, \tau) \qquad \text{and} \qquad \psi(\tau) = \varphi_u(\tau + \sigma)$$

for $x \in \mathbb{R}^n$, $\tau, \sigma \in \mathbb{R}$, then it follows that

$$\dot{\psi}(\tau) = f^u_\sigma(\psi(\tau), \tau) \qquad \text{for} \quad \tau \in \mathbb{R} \ \text{and} \ \psi(0) = \varphi(\sigma) = \varphi_i(x, \sigma).$$

This implies

$$\begin{aligned} \pi(x, f^u, \tau + \sigma) &= \left(\varphi_u(\tau + \sigma), f^u_{\tau+\sigma}\right) = \left(\psi(\tau), f^u_{\tau+\sigma}\right) \\ &= \pi\left(\varphi_u(x, \sigma), f^u_\sigma, \tau\right) = \pi\left(\pi(x, f^u, \sigma), \tau\right), \end{aligned}$$

i.e., the semigroup property also holds true.

For the proof of the continuity of $\pi : \mathbb{R}^n \times X \times \mathbb{R} \to \mathbb{R}^n \times X$ we refer to [33]. This shows that the mapping π defines a dynamical system in the sense of Section 1.1.

The Section 2.1.1–2.1.5 are taken from the book [22] and the Section 2.1.6 about time-minimal controllability of linear systems is taken from the book [13].

The Sections 2.2 and 2.3 about controllability of time-discrete dynamical systems are taken from the book [22] and Chapter 3 about chaotic behavior of dynamical systems can also be found in the book [14].

The paradigmatic example of a chaotic mapping is the shift-mapping in the space Σ of the infinite $0-1-$sequences for which not only the set of periodic points is dense in the space Σ but also there exists an orbit which is dense in Σ. From this one easily derives the topological transitivity (see Theorem 3.6). In [11] U. Kirchgraber calls the chaos that is generated by the shift-mapping "chaos in the sense of coin casting". This can be brought together with the chaotic behavior of one-dimensional continuous mappings $F : \mathbb{R} \to \mathbb{R}$ in the following way: Let us assume that there exist two closed finite intervals A_0 and A_1 with $A_0 \cap A_1 = \emptyset$,

$$F(A_0) \supseteq A_0 \cup A_1 \qquad \text{and} \qquad F(A_1) \supseteq A_0 \cup A_1.$$

Further we assume that $F : \mathbb{R} \to \mathbb{R}$ is differentiable and there exists a number $\lambda > 1$ such that

$$|F'(x)| \geq \lambda \qquad \text{for all} \quad x \in A_0 \cup A_1.$$

If we put, for every $k \in \mathbb{N}_0$,

$$A_{s_0 s_1 \ldots s_k} = \left\{ x \in A_{s_0} \,\middle|\, F(x) \in A_{s_1}, \ldots, F^k(x) \in A_{s_k} \right\}$$

where $s_j = 0$ or 1 for $j = 0, \ldots, k$, then it follows that

$$A_{s_0} \supseteq A_{s_0 s_1} \supseteq \ldots \supseteq A_{s_0 s_1 \ldots s_k}.$$

In [11] it is shown that

$$A_s = \bigcap_{k \in \mathbb{N}_0} A_{s_0 s_1 \ldots s_k}$$

is non-empty and consists of exactly one point $x = x(s) \in A_0 \cup A_1$, $s = (s_j)_{j \in \mathbb{N}_0}$. This gives rise to a mapping $h : \Sigma \to A_0 \cup A_1$, $\Sigma = \left\{ (s_j)_{j \in \mathbb{N}_0} \,\middle|\, s_j = 0 \text{ or } 1 \right\}$, with $h(s) = x(s)$ which can be shown to be continuous, if one introduces a metric in Σ, for instance, by defining

$$d(s, s') = 2^{-j} \qquad \text{where } j \text{ is the smallest index with } s_j \neq s'_j.$$

If one defines in $A_0 \cup A_1$ the set $A = h(\Sigma)$, then it follows that $F(A) \subseteq A$ and

$$F \circ h(s) = h \circ \sigma(s) \qquad \text{for all} \quad s \in \Sigma.$$

Theorem 3.7 therefore implies that $F : A \to A$ is chaotic in the sense of Devaney.

An example for this situation is the quadratic mapping

$$F(x) = \mu x(1-x), \; x \in \mathbb{R} \qquad \text{for} \quad \mu > 2 + \sqrt{5}$$

in Section 3.2 which also has been discussed in the book of Devaney.

The definition of topological entropy as a measure for chaos has been taken from the paper [28] by A. Mielke. In this paper also Theorem 3.16 is proved which states that the topological entropy is invariant under topological conjugacy.

The formulae which we have derived for the topological entropy of the shift-mapping σ on Σ and on a suitable subset Σ^A of Σ in Section 3.3 are a special case of a general formula for so called "subshifts" which is given in [28] in Theorem 2.1.

Also the connection with a one-dimensional mapping is given there a more general representation than in this book.

In Section 3.4 we have shown that disorder-chaos for one-dimensional mappings implies chaos in the sense of Li and Yorke. The reverse statement is false in general. J. Smital shows in [34] that there are chaotic mappings in the sense of Li and Yorke that have a vanishing topological entropy. In order to prove that the two chaos concepts are equivalent one has to sharpen the concept of Li and Yorke in a suitable way (see the paper [10] by Janková and Smital).

In Section 3.3.2 we have shown that the topological entropy of the shift-mapping $\sigma : \Sigma \to \Sigma$ is equal to $\ln 2$ and therefore positive so that it generates disorder chaos. This implies that every one-dimensional continuous mapping $f : J \to J$ of a finite closed interval J into itself is chaotic in the sense of Li and Yorke, if it is topologically conjugated with the shift-mapping $\sigma : \Sigma \to \Sigma$.

The representation of strange attractors in Section 3.5 is closely related to that of the book [5] of Devaney.

The Appendix A is the content of the paper [19].

A

A Dynamical Method for the Calculation of Nash-Equilibria in $n-$Person Games

A.1 The Fixed Point Theorems of Brouwer and Kakutani

The fixed point theorem of Kakutani is a generalization of Bouwer's fixed point theorem which says that every continuous mapping of a convex and compact subset of an $n-$dimensional Euclidean space into itself has at least one fixed point.

For the formulation of Kakutani's fixed point theorem we need the concept of set-valued mappings. For this purpose we consider a non-empty subset $K \subseteq \mathbb{R}^n$ (equipped with the Euclidean norm) and a mapping F of K into the power set 2^K of K which assigns to every point $x \in K$ a non-empty subset $F(x) \subseteq K$. A point $\widehat{x} \in K$ is called a fixed point of F, if $\widehat{x} \in F(\widehat{x})$ holds true. If in particular F is a point-map, i.e., $F(x)$, for every $x \in K$, is a one-point set, then this definition coincides with the normal definition of a fixed point.

Definition. *A mapping $F : K \to 2^K$ is called upper semi-continuous in $x \in K$, if for every sequence $(x_k)_{k\in\mathbb{N}}$ in K with $x_k \to x$ and every every sequence $(y_k)_{k\in\mathbb{N}}$ in K with $y_k \in F(x_k)$ for all $k \in \mathbb{N}$ and $y_k \to y$ it follows that $y \in F(x)$.*

If $F : K \to 2^K$ is a point-map, i.e., $F : K \to K$, then F is upper semi-continuous in $x \in K$, if and only if it is continuous in $x \in K$.

After these preparations we formulate the

Fixed Point Theorem of Kakutani: *Let $K \subseteq \mathbb{R}^n$ be a compact and convex subset and let $F : K \to 2^K$ be a mapping into the set of the compact and convex subsets of K which is upper semi-continuous in every $x \in K$, then F possesses a fixed point, i.e., there is some $\widehat{x} \in K$ with $\widehat{x} \in F(\widehat{x})$.*

S. Pickl and W. Krabs, *Dynamical Systems: Stability, Controllability and Chaotic Behavior*,
DOI 10.1007/978-3-642-13722-8_4,

If in particular F is a mapping of K into K which is continuous in every $x \in K$, then F has a fixed point $\widehat{x} \in K$ with $\widehat{x} = F(\widehat{x})$ which shows that the fixed point theorem of Kakutani implies the fixed point theorem of Brouwer.

For the proof of the fixed point theorem of Kakutani we refer to [4].

A.2 Nash Equilibria as Fixed Points of Mappings

The starting point of our investigation is a non-cooperative $n-$person game with strategy sets S_i, $i = 1, \ldots, n$, each for every player P_i, and pay-off functions $\Phi_i : \prod_{i=1}^{n} S_i = S \to \mathrm{I\!R}$ for $i = 1, \ldots, n$. Every player P_i endeavors to maximize the pay-off $\Phi_i(s)$, $s \in S$. This, however, is simultaneously in general impossible. So the players have to compromise. This gives rise to the following definition of a so called Nash equilibrium:

An $n-$tupel $(\widehat{s}_1, \ldots, \widehat{s}_n)$ of strategies $\widehat{s}_i \in S_i$ for $i = 1, \ldots, n$ is called a Nash equilibrium, if for all $i = 1, \ldots, n$ it is true that

$$\Phi_i \left(\widehat{s}_1, \ldots, \widehat{s}_n\right) \geq \Phi_i \left(\widehat{s}_1, \ldots, \widehat{s}_{i-1}, s_i, \widehat{s}_{i+1}, \ldots, \widehat{s}_n\right) \qquad \text{for all} \quad s_i \in S_i. \tag{A.1}$$

Interpretation: If a player deviates from his strategy in a Nash equilibrium whereas the other players stick to their strategies, his pay-off can at most diminish.

In the following, we are concerned with finding a method to calculate Nash equilibria. The general idea behind this is to find a mapping $f : S \to S$ such that every fixed point $\widehat{s} \in S$ with $f(\widehat{s}) = \widehat{s}$ is a Nash equilibrium. If such a mapping has been found, it is natural to perform an iteration method by choosing $s_0 \in S$ and defining a sequence $(s_k)_{k \in \mathrm{I\!N}_0}$ in S via

$$s_{k+1} = f(s_k) = f^{k+1}(s_0), \qquad k \in \mathrm{I\!N}_0. \tag{A.2}$$

If this sequence converges to some $\widehat{s} \in S$ and if $f : S \to S$ is continuous, then it follows that $\widehat{s} = f(\widehat{s})$, i.e., $\widehat{s}$ is a fixed point of f and therefore a Nash equilibrium. The definition of a Nash equilibrium can be used in order to find a function of this kind. For that purpose we make the following assumption:

For every $n-$tupel $(\widehat{s}_1^*, \ldots, \widehat{s}_n^*) \in S$ and every $i = 1, \ldots, n$ there is exactly one $\widetilde{s}_i \in S_i$ with

$$\Phi_i \left(s_1^*, \ldots, s_{i-1}^*, \widetilde{s}_i, s_{i+1}^*, \ldots, s_n^*\right) \geq \Phi_i \left(s_1^*, \ldots, s_{i-1}^*, s_i, s_{i+1}^*, \ldots, s_n^*\right) \quad \text{for all} \quad s_i \in S_i. \tag{A.3}$$

Under this assumption we then define a mapping $f = (f_1, \ldots, f_n) : S \to S$ via

$$f_i(s_1^*, \ldots, s_n^*) = \widetilde{s}_i \qquad \text{for} \quad i = 1, \ldots, n \tag{A.4}$$

with $\widetilde{s}_i \in S_i$ being the unique solution of (A.3). It is obvious that $\widehat{s} \in S$ is a fixed point of f, if and only if $\widehat{s}$ is a Nash equilibrium.

In [22] we have shown that f is a continuous mapping of S into S, if in addition to the above assumption the following two assumptions are satisfied:

(1) The strategy sets S_i, $i = 1, \ldots, n$, are convex and compact subsets of some $\mathbb{R}^{m_i}$.
(2) The pay-off functions $\Phi_i : S \to \mathbb{R}$ are continuous.

Since the set $S \subseteq \mathbb{R}^{m_1+\ldots+m_n}$ is convex and compact, Brouwer's fixed point theorem implies that f has at least one fixed point. Further the assumptions (1) and (2) guarantee that, for every $n-$tupel $s^* \in S$, there exists an $n-$tupel $\widetilde{s} \in S$ such that (A.3) is satisfied. This shows that the iteration (A.2) can be performed without the uniqueness assumption on $\widetilde{s} \in S$.

In this case we define the set-valued mapping f of S into the power set 2^S of S which assigns to every $s^* \in S$ the set $f(s^*)$ of all $\widetilde{s} \in S$ such that (A.3) is satisfied.

Obviously it follows that $\widehat{s} \in S$ is a Nash equilibrium, if and only if $\widehat{s} \in f(\widehat{s})$. By the assumptions (1) and (2), for every $s^* \in S$ the set $f(s^*)$ is non-empty and compact.

Now let $\left(s^k\right)_{k\in\mathbb{N}}$ and $\left(\widetilde{s}^k\right)_{k\in\mathbb{N}}$ be sequences with

$$\widetilde{s}^k \in f\left(s^k\right), \qquad s^* = \lim_{k\to\infty} s^k$$

and

$$\widetilde{s} = \lim_{k\to\infty} \widetilde{s}^k \qquad \text{for some} \quad s^* \in S \quad \text{and} \quad \widetilde{s} \in S.$$

Then we have, for every $k \in \mathbb{N}$

$$\Phi_i\left(s_1^k, \ldots, s_{i-1}^k, \widetilde{s}_i^k, s_{i+1}^k, \ldots, s_n^k\right) \geq \Phi_i\left(s_1^k, \ldots, s_{i-1}^k, s_i, s_{i+1}^k, \ldots, s_n^k\right)$$
$$\text{for all} \quad s_i \in S_i \quad \text{and} \quad i = 1, \ldots, n.$$

This implies

$$\Phi_i\left(s_1^*, \ldots, s_{i-1}^*, \widetilde{s}_i, s_{i+1}^*, \ldots, s_n^*\right) \geq \Phi_i\left(s_1^*, \ldots, s_{i-1}^*, s_i, s_{i+1}^*, \ldots, s_n^*\right)$$
$$\text{for all} \quad s_i \in S_i \quad \text{and} \quad i = 1, \ldots, n,$$

i.e., $\widetilde{s} \in F(s^*)$. This shows that the mapping f is upper semi-continuous in every $s^* \in S$.

If in addition to the assumptions (1) and (2) we assume that, for every $s^* \in S$, the functions $\Phi_i\left(s_1^*, \ldots, s_{i-1}^*, \cdot, s_{i+1}^*, \ldots, s_n^*\right)$ are concave, then for every $s^* \in S$ the set $f(s^*)$ is convex.

Summarizing we conclude that under the assumptions (1) and (2) and the last assumption the mapping $f : S \to 2^S$ defined by (A.3) maps S into the set of convex and compact subsets of S and is upper semi-continuous in every $s^* \in S$.

By Kakutani's fixed point theorem it follows that f has a fixed point, i.e., there exists some $\widehat{s} \in S$ with $\widehat{s} \in f(\widehat{s})$ which is a Nash equilibrium.

The sequence $(s_k)_{k\in\mathbb{N}_0}$ defined by (A.2) now has to be replaced by a sequence $\left(s^k\right)_{k\in\mathbb{N}_0}$ in S which is defined as follows:

If $s^k \in S$ is given for some $k \in \mathbb{N}_0$, then we determine, for every $i = 1,\ldots,n$, some $s_i^{k+1} \in S_i$ with

$$\begin{aligned}\Phi_i\left(s_1^k,\ldots,s_{i-1}^k,s_i^{k+1},s_{i+1}^k,\ldots,s_n^k\right)\\ \geq \Phi_i\left(s_1^k,\ldots,s_{i-1}^k,s_i,s_{i+1}^k,\ldots,s_n^k\right) \qquad \text{for all} \quad s_i \in S_i. \qquad \text{(A.5)}\end{aligned}$$

and put $s^{k+1} = \left(s_1^{k+1},\ldots,s_n^{k+1}\right)$.

If then there exists some $\widehat{s} \in S$ such that $\widehat{s} = \lim\limits_{k\to\infty} s^k$, it follows from (A.5) that (A.1) is satisfied, i.e., $\widehat{s}$ is a Nash equilibrium.

Now we assume in addition to the assumptions (1) and (2) that for every $s^* \in S$ the functions $\Phi_i\left(s_1^*,\ldots,s_{i-1}^*,\cdot,s_{i+1}^*,\ldots,s_n^*\right) : S_i \to \mathbb{R}$, $i = 1,\ldots,n$, are concave and Gateaux-differentiable.

Then $\widehat{s} \in S$ is a Nash equilibrium, if and only if

$$\nabla_{s_i}\Phi_i\left(\widehat{s}\right)^T \widehat{s}_i \geq \nabla_{s_i}\Phi_i\left(\widehat{s}\right)^T s_i \qquad \text{for all} \quad s_i \in S_i \quad \text{and} \quad i = 1,\ldots,n \quad \text{(A.6)}$$

where $\nabla_{s_i}\Phi_i(\cdot)$ is the gradient of Φ_i with respect to s_i.
The iteration procedure (A.5) can therefore be replaced by the following:

If $s^k \in S$ is given for some $k \in \mathbb{N}_0$, then we determine, for every $i = 1,\ldots,n$, some $s_i^{k+1} \in S_i$ with

$$\nabla_{s_i}\Phi_i\left(s^k\right)^T s_i^{k+1} \geq \nabla_{s_i}\Phi_i\left(s^k\right)^T s_i \qquad \text{for all} \quad s_i \in S_i \qquad \text{(A.7)}$$

and put $s^{k+1} = \left(s_1^{k+1},\ldots,s_n^{k+1}\right)$.

If, for every $i = 1,\ldots,n$, the mapping $\nabla_{s_i}\Phi_i : S \to \mathbb{R}^{m_i}$ is continuous and if there exists some $\widehat{s} = \lim\limits_{k\to\infty} s^k$, $\widehat{s} \in S$, then it follows from (A.7) that (A.6) is satisfied, i.e., $\widehat{s}$ is a Nash equilibrium.

Special cases:

(a) Let, for every $i = 1,\ldots,n$,

$$S_i = \left\{ s_i \in \mathbb{R}^{m_i} \;\middle|\; s_{ij} \geq 0 \text{ for } j = 1,\ldots,m_i \text{ and } \sum_{j=1}^{m_i} s_{ij} = 1 \right\}.$$

Then, for every $s \in S$, it follows that

$$\nabla_{s_i}\Phi_i(s)^T s_i \leq \max_{j=1,\ldots,m_i} \Phi_{i s_{ij}}(s) \qquad \text{for} \quad i = 1,\ldots,n.$$

This implies that $\widehat{s} \in S$ is a Nash equilibrium, if and only if

$$\nabla_{s_i} \Phi_i(\widehat{s})^T \widehat{s}_i = \max_{j=1,\ldots,m_i} \Phi_{is_{ij}}(\widehat{s}) \qquad \text{for} \quad i=1,\ldots,n.$$

Now we define, for every $s \in S$,

$$\varphi_{ij}(s) = \max\left(0, \Phi_{is_{ij}}(s) - \nabla_{s_i}\Phi_i(s)^T s_i\right), \quad j=1,\ldots,m_i, \quad i=1,\ldots,n.$$

Then it follows that $\widehat{s} \in S$ is a Nash equilibrium, if and only if

$$\varphi_{ij}(\widehat{s}) = 0 \qquad \text{for all} \quad j=1,\ldots,m_i \quad \text{and} \quad i=1,\ldots,n. \tag{A.8}$$

Now we define a mapping $f=(f_1,\ldots,f_n) : S \to S$ via

$$f_i(s)_k = \frac{1}{1+\sum\limits_{j=1}^{m_i}\varphi_{ij}(s)}\left(s_{ik} + \underbrace{\sum_{j=1}^{m_i}\varphi_{ij}(s)\,e_j^k}_{\varphi_{ik}(s)}\right), \quad k=1,\ldots,m_i,\ i=1,\ldots,n, \tag{A.9}$$

where $e_j^k = \begin{Bmatrix} 0 & \text{for} & k \neq j, \\ 1 & \text{for} & k=j. \end{Bmatrix}$. Since f is a continuous mapping from S into S. Brouwer's fixed point theorem implies that there exists a fixed point $\widehat{s} \in S$ of f.

Assertion: $\widehat{s} \in S$ is a fixed point of f, i.e., $\widehat{s} = f(\widehat{s})$, if and only if (A.8) holds true.

Proof. 1) If (A.8) holds true for some $\widehat{s} \in S$, then $\widehat{s}$ is a fixed point of f.

2) Conversely let $\widehat{s} = f(\widehat{s})$. For every $i=1,\ldots,n$ there exists at least one $k \in \{1,\ldots,m_i\}$ with $\widehat{s}_{ik} > 0$. We choose $k_0 \in \{1,\ldots,m_i\}$ such that

$$\Phi_{is_{ik_0}}(\widehat{s}) = \min\left\{\Phi_{is_{ik}}(\widehat{s}) \mid \widehat{s}_{ik} > 0\right\} \quad \text{and conclude that} \quad \varphi_{ik_0}(\widehat{s}) = 0.$$

This, however, implies that $\varphi_{ik}(\widehat{s}) = 0$ for all $k=1,\ldots,m_i$, since otherwise $\widehat{s}_{ik_0} < \widehat{s}_{ik_0}$ would follow which is impossible.

Therefore, if one defines, starting with some $s^0 \in S$, a sequence $\left(s^k\right)_{k\in\mathbb{N}_0}$ in S via $s^{k+1} = f\left(s^k\right)$ and does this sequence converge to some $\widehat{s} \in S$, then $\widehat{s}$ is a fixed point of f and hence a Nash equilibrium. □

Remark. *The construction of the mapping f is the same as that which has been given by J. Nash in [31] in order to prove the existence of Nash equilibria for the mixed extension of an $n-$person game with finite strategy sets.*

(b) Let, for every $i = 1, \ldots, n$,

$$S_i = \{s_i \in \mathbb{R}^{m_i} | \ ||s_i||_2 \leq 1\}$$

where $|| \cdot ||_2$ denotes the Euclidean norm.

Then we define a mapping $f = (f_1, \ldots, f_n) : S \to S$ via

$$f_i(s) = \left\{ \begin{array}{ll} \Theta_{m_i} & \text{if} \quad \nabla_{s_i} \Phi_i(s) = \Theta_{m_i}, \\ \dfrac{\nabla_{s_i} \Phi_i(s)}{||\nabla_{s_i} \Phi_i(s)||_2} & \text{if} \quad \nabla_{s_i} \Phi_i(s) \neq \Theta_{m_i}. \end{array} \right\} \quad \text{for} \quad i = 1, \ldots, n.$$

This mapping is continuous. Therefore Brouwer's fixed point theorem implies that f has at least one fixed point $\widehat{s} = f(\widehat{s})$ and for every such it follows

$$\nabla_{s_i} \Phi_i (\widehat{s})^T \ s_i \leq ||\nabla_{s_i} \Phi_i (\widehat{s}) \, ||_2 = \nabla_{s_i} \Phi_i (\widehat{s})^T \ \widehat{s}_i$$
$$\text{for all} \quad s_i \in S_i \quad \text{and} \quad i = 1, \ldots, n.$$

which shows that $\widehat{s}$ is a Nash equilibrium.

Therefore, if one defines, starting with some $s^0 \in S$, a sequence $\left(s^k\right)_{k \in \mathbb{N}_0}$ in S via $s^{k+1} = f\left(s^k\right)$ and does this sequence converge to some $\widehat{s} \in S$, then $\widehat{s}$ is a fixed point of f and hence a Nash equilibrium.

A.3 Bi-Matrix Games

We consider 2−person games with finite strategy sets which are played with mixed strategies so that their strategy sets are given by

$$S_1 = \left\{ s \in \mathbb{R}^m \,\middle|\, s_j \geq 0 \;\text{ for }\; j = 1, \ldots, m \;\text{ and }\; \sum_{j=1}^{m} s_j = 1 \right\}$$

and

$$S_2 = \left\{ t \in \mathbb{R}^n \,\middle|\, t_k \geq 0 \;\text{ for }\; k = 1, \ldots, n \;\text{ and }\; \sum_{k=1}^{n} t_k = 1 \right\}.$$

Thus S_1 and S_2 are convex and compact in $\mathbb{R}^m$ and $\mathbb{R}^n$, respectively.

The pay-off functions are given by

$$\Phi_1(s,t) = s^T A t \quad \text{and} \quad \Phi_2(s,t) = s^T B t \qquad \text{for} \quad (s,t) \in S = S_1 \times S_2$$

where A and B are given $m \times n$−matrices.

Obviously Φ_1 and Φ_2 are continuous on S and for every $(s^*, t^*) \in S$ the functions $\Phi_1(s, t^*) = s^T A t^*$, $s \in S_1$, and $\Phi_2(s^*, t) = s^{*T} A t$, $t \in S_2$, are linear, hence concave, and Gateaux-differentiable with gradients

$$\nabla \Phi_1(s, t^*) = A t^*, \; s \in S_1 \qquad \text{and} \qquad \nabla \Phi_2(s^*, t) = B s^*, \; t \in S_2,$$

which are continuous as functions on S.

So we are in the situation of special case (a) in Section A.3. the functions $\Phi_{ij} : S \to \mathbb{R}$ defined there for $j = 1, \ldots, m_i$ and $i = 1, \ldots, m$ are here given by

$$\varphi_{1j}(s,t) = \max\left(0, e_j^T A t - s^T A t\right) \qquad \text{for} \quad j = 1, \ldots, m$$

and

$$\varphi_{2k}(s,t) = \max\left(0, s^T B e_k - s^T B t\right) \qquad \text{for} \quad k = 1, \ldots, n$$

where e_j and e_k is the j−th and the k−th unit vector in $\mathbb{R}^m$ and $\mathbb{R}^n$, respectively.

Further $(\widehat{s}, \widehat{t}) \in S$ is a Nash equilibrium, if and only if

$$\varphi_{1j}\left(\widehat{s}, \widehat{t}\right) = 0, \; j = 1, \ldots, m \qquad \text{and} \qquad \varphi_{2k}\left(\widehat{s}, \widehat{t}\right) = 0, \; k = 1, \ldots, n. \quad \text{(A.10)}$$

The mapping f defined by (A.9) here reads $f = (f_1, f_2) : S \to S$ where

$$f_1(s,t)_j = \frac{s_j + \varphi_{1j}(s,t)}{1 + \sum_{i=1}^{m} \varphi_{1i}(s,t)}, \qquad j = 1, \ldots, m,$$

and

$$f_2(s,t)_k = \frac{t_k + \varphi_{2k}(s,t)}{1 + \sum_{j=1}^{n} \varphi_{2j}(s,t)}, \qquad k = 1, \ldots, n.$$

As in Section A.3 this mapping has a fixed point $(\widehat{s}, \widehat{t}) \in S$ which must satisfy (A.10) and therefore is a Nash equilibrium (see Section A.3).

An example: Let $m = 2,\ n = 3$ and

$$A = \begin{pmatrix} 5 & 3 & 1 \\ 6 & 4 & 2 \end{pmatrix}, \qquad B = \begin{pmatrix} 4 & 5 & 6 \\ 3 & 2 & 1 \end{pmatrix}.$$

Then we obtain

$$\begin{aligned} e_1^T At &= 5t_1 + 3t_2 + t_3, \\ e_2^T At &= 6t_1 + 4t_2 + 2t_3, \\ e_3^T At &= 6t_1 + 4t_2 + 2t_3 - s_1, \end{aligned}$$

hence

$$e_1^T At - s^T At = -1 + s_1 \le 0,$$

and

$$e_2^T At - s^T At = s_1 > 0, \qquad \text{if} \quad s_1 > 0.$$

This implies

$$\begin{gathered} \varphi_{11}(s,t) = 0 \qquad \text{and} \qquad \varphi_{12}(s,t) = s_1 > 0 \\ \text{for all} \quad (s,t) \in S \quad \text{with} \quad 0 < s_1 \le 1 \end{gathered}$$

and further

$$f_1(s,t)_1 = \frac{s_1}{1 + s_1}, \qquad f_1(s,t)_2 = \frac{1}{1 + s_1} = 1 - f_1(s,t)_1.$$

Next we obtain

$$
\begin{aligned}
s^T B e_1 &= 4s_1 + 3s_2 = 3 + s_1, \\
s^T B e_2 &= 5s_1 + 2s_2 = 2 + 3s_1, \\
s^T B e_3 &= 6s_1 + s_2 = 1 + 5s_1, \\
s^T B t &= 3t_1 + 2t_2 + t_3 + s_1 (t_1 + 3t_2 + 5t_3)
\end{aligned}
$$

which implies

$$
\begin{aligned}
s^T B e_1 - s^T B t &= 3 - s_1 (2t_2 + 4t_3) - 3t_1 - 2t_2 - t_3 \\
&= 3 - 3t_1 - 2 (s_1 + 1) t_2 - (4s_1 + 1) t_3 \\
&> 0 \qquad \text{if} \quad s_1 < \frac{1}{2}, \\
s^T B e_2 - s^T B t &= 2 + s_1 (2t_1 - 2t_3) - 3t_1 - 2t_2 - t_3 \\
&< 2 + t_1 - t_3 - 3t_1 - 2t_2 - t_3 \\
&= 0 \qquad \text{if} \quad t_1 > t_3 \quad \text{and} \quad s_1 < \frac{1}{2}, \\
s^T B e_3 - s^T B t &= 1 + s_1 (4t_1 + 2t_2) - 3t_1 - 2t_2 - t_3 \\
&< 1 + 2t_1 + t_2 - 3t_1 - 2t_2 - t_3 \\
&= 0 \qquad \text{if} \quad s_1 < \frac{1}{2}.
\end{aligned}
$$

Further

$$
\begin{aligned}
&\varphi_{21}(s,t) > 0 \qquad \text{and} \qquad \varphi_{22}(s,t) = \varphi_{23}(s,t) = 0 \\
&\text{for all} \quad (s,t) \in S \quad \text{with} \quad s_1 < \frac{1}{2} \quad \text{and} \quad t_1 > t_3.
\end{aligned}
$$

We therefore have

$$
\begin{aligned}
f_2(s,t)_1 &= \frac{t_1 + \varphi_{21}(s,t)}{1 + \varphi_{21}(s,t)} = 1 - f_2(s,t)_2 - f_2(s,t)_3, \\
f_2(s,t)_2 &= \frac{t_2}{1 + \varphi_{21}(s,t)}, \\
f_2(s,t)_3 &= \frac{t_3}{1 + \varphi_{21}(s,t)} \qquad \text{for all} \quad (s,t) \in S \quad \text{with} \quad s_1 < \frac{1}{2} \quad \text{and} \quad t_1 > t_3.
\end{aligned}
$$

Hence, if one chooses $(s^0, t^0) \in U = \{(s,t) \in S \mid 0 < s_1 < 1/2 \text{ and } t_1 > t_3\}$ and defines a sequence $(s^k, t^k)_{k \in \mathbb{N}_0}$ via $(s^{k+1}, t^{k+1}) = f(s^k, t^k)$ then because of

$$\begin{gathered} f_1(s,t)_1 < s_1, \quad f_2(s,t)_2 < t_2, \quad f_2(s,t)_3 < t_3 \\ \Downarrow \\ f_2(s,t)_1 = 1 - f_2(s,t)_2 - f_2(s,t)_3 > 1 - t_2 - t_3 = t_1 \end{gathered}$$

for all $(s,t) \in U$ it follows that

$$(s^k, t^k) \in U \qquad \text{for all} \quad k \in \mathbb{N}_0.$$

Further it follows that

$$s_1^k \to 0, \quad s_2^k \to 1, \quad t_1^k \to 1, \quad t_2^k \to 0, \quad t_3^k \to 0,$$

hence $(s^k, t^k) \to (e_2, e_1) \in \overline{U}$ and (e_2, e_1) is a Nash equilibrium.

This example is a special case of the following general situation:

Let $U \subseteq S$ be such that there exists some $j_0 \in \{1, \ldots, m\}$ and some $k_0 \in \{1, \ldots, n\}$ with

$$\left.\begin{array}{l} \varphi_{1j}(s,t) = 0 \text{ for all } j \neq j_0, \ \varphi_{1j_0}(s,t) > 0, \\ \varphi_{2k}(s,t) = 0 \text{ for all } k \neq k_0, \ \varphi_{2k_0}(s,t) > 0 \\ \qquad\qquad\qquad\qquad \text{for all} \quad (s,t) \in U. \end{array}\right\}$$

Then it follows that

$$\begin{aligned} f_1(s,t)_j &= \frac{s_j}{1 + \varphi_{1j_0}(s,t)} \qquad \text{for all} \quad j \neq j_0, \\ f_1(s,t)_{j_0} &= \frac{s_{j_0} + \varphi_{1j_0}(s,t)}{1 + \varphi_{1j_0}(s,t)} = 1 - \sum_{j \neq j_0} f_1(s,t)_j, \\ f_2(s,t)_k &= \frac{t_k}{1 + \varphi_{2k_0}(s,t)} \qquad \text{for all} \quad k \neq k_0, \\ f_2(s,t)_{k_0} &= \frac{t_{k_0} + \varphi_{2k_0}(s,t)}{1 + \varphi_{2k_0}(s,t)} = 1 - \sum_{k \neq k_0} f_2(s,t)_k. \end{aligned}$$

This implies

$$\left.\begin{array}{l} f_1(s,t)_j < s_j \text{ for all } j \neq j_0 \Rightarrow f_1(s,t)_{j_0} > s_{j_0}, \\ f_2(s,t)_k < t_k \text{ for all } k \neq k_0 \Rightarrow f_2(s,t)_{k_0} > t_{k_0} \\ \qquad\qquad\qquad\qquad \text{for all} \quad (s,t) \in U. \end{array}\right\} \tag{A.11}$$

Assumption: $f(U) \subseteq U$.

If we now define, starting with some $(s^0, t^0) \in U$, a sequence $((s^k, t^k))_{k \in \mathbb{N}_0}$ in U via $s^{k+1} = f_1(s^k, t^k)$, $t^{k+1} = f_2(s^k, t^k)$, then it follows from (A.11) that $s^k \to \widehat{s}$ and $t^k \to \widehat{t}$ with $(\widehat{s}, \widehat{t}) = f(\widehat{s}, \widehat{t}) \in \overline{U}$.

This implies that $(\widehat{s}, \widehat{t})$ is a Nash equilibrium.

From this it follows in particular that

$$\begin{aligned} e_{j_0}^T A\widehat{t} - s^T A\widehat{t} &= \sum_{k=1}^{n} a_{j_0 k}\,\widehat{t}_k - \sum_{j=1}^{m} \widehat{s}_j \sum_{k=1}^{n} a_{jk}\,\widehat{t}_k \\ &= \sum_{j=1}^{m} \widehat{s}_j \left(\sum_{k=1}^{n} a_{j_0 k}\,\widehat{t}_k - \sum_{k=1}^{n} a_{jk}\,\widehat{t}_k \right) \\ &= \sum_{\substack{j=1 \\ j \neq j_0}}^{m} \widehat{s}_j \sum_{k=1}^{n} (a_{j_0 k} - a_{jk})\,\widehat{t}_k \leq 0. \end{aligned}$$

Assumption 1:

$$a_{j_0 k} - a_{jk} > 0 \quad \text{for all} \quad j \neq j_0 \quad \text{and} \quad k = 1, \ldots, n.$$

Then it follows that

$$\widehat{s}_j = 0 \quad \text{for all} \quad j \neq j_0, \qquad \text{hence} \quad \widehat{s}_{j_0} = 1 \quad \text{and therefore} \quad \widehat{s} = e_{j_0}.$$

This in turn implies

$$e_{j_0}^T B e_{k_0} - e_{j_0}^T B\widehat{t} = b_{j_0 k_0} - \sum_{k=1}^{n} b_{j_0 k}\,\widehat{t}_k = \sum_{\substack{k=1 \\ k \neq k_0}}^{n} (b_{j_0 k_0} - b_{j_0 k})\,\widehat{t}_k \leq 0.$$

Assumption 2:

$$b_{j_0 k_0} - b_{j_0 k} > 0 \quad \text{for all} \quad k \neq k_0.$$

Then it follows that

$$\widehat{t}_k = 0 \quad \text{for all} \quad k \neq k_0, \qquad \text{hence} \quad \widehat{t}_{k_0} = 1 \quad \text{and therefore} \quad \widehat{t} = e_{k_0}.$$

In the above example Assumption 1 is satisfied for $j_0 = 2$ and Assumption 2 for $j_0 = 2$ and $k_0 = 1$.

Next we consider symmetric bi-matrix games in which $m = n$, hence $S_1 = S_2 = S = \left\{ s \in \mathbb{R}^n \mid s_j \geq 0 \text{ and } \sum_{j=1}^{n} s_j = 1 \right\}$, and $A = B^T$. Then, one can show that there exists a Nash equilibrium of the form $(\widehat{s}, \widehat{s})$ with $\widehat{s} \in S$ which is characterized by the condition

$$\widehat{s}^T A\widehat{s} = \max_{j=1,\ldots,n} e_j^T A\widehat{s}. \tag{A.12}$$

If we define functions $\varphi_j : S \to \mathbb{R}$ by

$$\varphi_j(s) = \max\left(0, e_j^T A\, s - s^T A\, s\right) \qquad \text{for} \quad j = 1, \ldots, n,$$

then condition (A.12) is equivalent to

$$\varphi_j\left(\widehat{s}\right) = 0 \qquad \text{for all} \quad j = 1, \ldots, n. \tag{A.13}$$

We further define a mapping $f = (f_1, \ldots, f_n) : S \to S$ via

$$f_j(s) = \frac{1}{1 + \sum\limits_{k=1}^{n} \varphi_k(s)} \left(s_j + \varphi_j(s)\right) \qquad \text{for} \quad j = 1, \ldots, n, \tag{A.14}$$

then condition (A.13) turns out to be equivalent to $f\left(\widehat{s}\right) = \widehat{s}$ (see Section A.3). Thus $\left(\widehat{s}, \widehat{s}\right)$, $\widehat{s} \in S$ is a Nash equilibrium, if and only if $\widehat{s}$ is a fixed point of the mapping $f = (f_1, \ldots, f_n)$ defined by (A.14).

An example: Let $m = n = 3$ and

$$A = \begin{pmatrix} 7 & 5 & 3 \\ 8 & 4 & 2 \\ 9 & 5 & 1 \end{pmatrix} = B^T.$$

Then we obtain, for every $s \in S$,

$$\begin{aligned}
e_1^T A\, s &= 7s_1 + 5s_2 + 3s_3, \\
e_2^T A\, s &= 8s_1 + 4s_2 + 2s_3, \\
e_3^T A\, s &= 9s_1 + 5s_2 + s_3, \\
s^T A\, s &= s_1\left(7s_1 + 5s_2 + 3s_3\right) + s_2\left(8s_1 + 4s_2 + 2s_3\right) + s_3\left(9s_1 + 5s_2 + s_3\right)
\end{aligned}$$

which implies

$$\left.\begin{aligned}
e_1^T A\, s - s^T A\, s &= s_2^2, \\
e_2^T A\, s - s^T A\, s &= s_2^2 - s_2, \\
e_3^T A\, s - s^T A\, s &= s_2^2,
\end{aligned}\right\} \qquad \text{if} \quad s_1 = s_3.$$

From this it follows that

$$\varphi_1(s) = \varphi_3(s) = s_2^2 > 0 \quad \text{and} \quad \varphi_2(s) = 0 \qquad \text{for all} \quad s \in U$$

with

$$U = \{s \in S \mid s_1 = s_3,\ s_2 > 0\}.$$

The functions (A.14) for $s \in U$ are given by

$$f_1(s) = \frac{s_1 + s_2^2}{1 + 2s_2^2}, \quad f_2(s) = \frac{s_2}{1 + 2s_2^2}, \quad f_3(s) = \frac{s_1 + s_2^2}{1 + 2s_2^2}.$$

This shows that $f(U) \subseteq U$.

If we now define, starting with some $s^0 \in U$, a sequence $\left(s^k\right)_{k\in\mathbb{N}_0}$ in U via $s^{k+1} = f\left(s^k\right),\ k \in \mathbb{N}_0$, then it follows from

$$f_2(s) < s_2 \qquad \text{for all} \quad s \in U$$

that $s_2^k \to \widehat{s}_2 \geq 0$ and $\dfrac{\widehat{s}_2}{1+2\widehat{s}_2} = \widehat{s}_2 \;\Leftrightarrow\; 2\widehat{s}_2^2 = 0 \;\Leftrightarrow\; \widehat{s}_2 = 0.$

Because of $f_1(s) = (1 - f_2(s))/2$ this implies

$$s_1^{k+1} = f_1\left(s^k\right) = \frac{1}{2}\left(1 - f_2\left(s^k\right)\right) = \frac{1}{2}\left(1 - s_2^{k+1}\right) \to \frac{1}{2}.$$

Hence $s^k \to \widehat{s} = \left(\frac{1}{2}, 0, \frac{1}{2}\right)^T$ and $(\widehat{s}, \widehat{s})$ is a Nash equilibrium.

A.4 Evolution Matrix Games

An evolution game (see [26]) can be considered as a 1−person game whose player is a population of individuals that have a finite number of strategies $I_1, \ldots, I_n$ $(n > 1)$ in order to survive in the struggle of life. Let $s_i \in [0,1]$, for every $i = 1, \ldots, n$, be the probability for the strategy I_i to be chosen in the population. Then the corresponding state of the population is defined by the vector $s = (s_1, \ldots, s_n)$ where $\sum_{i=1}^{n} s_i = 1$. The space of all population states is given by the simplex

$$S = \left\{ s = (s_1, \ldots, s_n) \;\middle|\; 0 \leq s_i \leq 1,\ \sum_{i=1}^{n} s_i = 1 \right\}.$$

Every vector $e_i = (0, \ldots, 0, 1_i, 0, \ldots, 0)$, $i = 1, \ldots, n$, denotes a so called pure population state where all individuals choose the strategy I_i. All the other states are called mixed states. If an individual that chooses strategy I_i meets an individual that chooses strategy I_j, we assume that the I_i−individual is given a pay-off $a_{ij} \in \mathbb{R}$ by the I_j−individual. All the pay-offs then form a matrix

$$A = (a_{ij})_{i,j=1,\ldots,n},$$

the so called pay-off matrix which defines a matrix game.

The expected pay-off of an I_i−individual in the population state $s \in S$ is defined by

$$\sum_{j=1}^{n} a_{ij}\, s_j = e_i\, A t^T.$$

Definition. *A population state $s^* \in S$ is called a Nash equilibrium, if*

$$s\, A\, s^{*T} \leq s^*\, A\, s^{*T} \qquad \text{for all} \quad s \in S. \tag{A.15}$$

In words this means that the average pay-off of s^* to s^* is always not smaller than its average pay-off to any other state $s \in S$.

It is easy to see that condition (A.15) is equivalent to

$$s^*\, A\, s^{*T} = \max_{i=1,\ldots,n} e_i\, A\, s^{*T}. \tag{A.16}$$

Now let us define, for every $i = 1, \ldots, n$ and $s \in S$,

$$\varphi_i(s) = \max\left(0, e_i\, A\, s^T - s\, A\, s^T\right)$$

and

$$f_i(s) = \frac{1}{1 + \sum_{j=1}^{n} \varphi_j(s)} \left(s_i + \varphi_i(s)\right).$$

Then we obtain a continuous mapping $f = (f_1, \ldots, f_n) : S \to S$ which, by Brouwer's fixed point theorem, has at least one fixed point $s^* \in S$, i.e., $f(s^*) = s^*$. This implies

$$\varphi_i(s^*) = 0 \qquad \text{for all} \quad i = 1, \ldots, n \quad \text{(see Section A.3)}$$

and in turn (A.16) which shows that, if $s^* \in S$ is a fixed point of S (existence being insured), s^* is a Nash equilibrium.

Condition (A.16) implies

$$e_i\, A\, s^{*T} = s^*\, A\, s^{*T} \qquad \text{for all} \quad i \in \{1, \ldots, n\} \quad \text{with} \quad s_i^* > 0.$$

Now let J be a non-empty subset of $\{1, \ldots, n\}$ such that

$$e_i\, A\, s^{*T} < s^*\, A\, s^{*T} \qquad \text{for all} \quad i \in J.$$

Further, there exists an open subset $U \subseteq S$ with $s^* \in U$ such that

$$e_i\, A\, s^T < s\, A\, s^T \qquad \text{for all} \quad i \in J \quad \text{and all} \quad s \in U$$

which implies

$$\varphi_i(s) = 0 \qquad \text{for all} \quad i \in J \quad \text{and all} \quad s \in U,$$

hence

$$f_i(s) = \frac{s_i}{1 + \sum\limits_{j \notin J} \varphi_j(s)} \qquad \text{for all} \quad i \in J \quad \text{and all} \quad s \in U.$$

Now let us assume that for every $s \in U$ there is some $i \in J$ with $s_i > 0$. Further, we assume that for every $s \in U$ with $s \neq s^*$ there is some $j \notin J$ with $\varphi_j(s) > 0$. Then it follows that

$$\begin{aligned} f_i(s) &\le s_i \qquad \text{for all } i \in J \quad \text{and all} \quad s \in U \quad \text{with} \quad s \neq s^* \quad \text{and} \\ f_i(s) &< s_i \qquad \text{for some} \quad i = i(s) \in J. \end{aligned}$$

If we define a Lyapunov function $V : U \to \mathbb{R}$ by

$$V(s) = \sum_{i \in J} s_i, \qquad s \in U,$$

then it follows that

$$V(s^*) = 0 \quad \text{and} \quad V(s) > 0 \qquad \text{for all} \quad s \in U \quad \text{with} \quad s \neq s^*$$

and

$$V(f(s)) = \sum_{i \in J} f_i(s) < \sum_{i \in J} s_i = V(s) \qquad \text{for all} \quad s \in U,\ s \neq s^*.$$

This implies that s^* is an asymptotically stable fixed point of f (see Theorem 1.3).

This means in particular that s^* is an attractor, i.e., there exists an open subset $W \subseteq S$ with $s^* \in W$ such that for every $s^0 \in W$ the sequence $\left(s^k\right)_{k \in \mathbb{N}_0}$ defined by $s^{k+1} = f\left(s^k\right),\ k \in \mathbb{N}_0$, converges to s^*.

Now let $U \subseteq S$ be such that

$$\varphi_j(s) = 0 \qquad \text{for all} \quad j \in J \subseteq \{1, \ldots, n\},\ |J| < n \quad \text{and all} \quad s \in U.$$

Then it follows, for every $s \in U$, that

$$f_i(s) = \frac{s_i}{1 + \sum\limits_{j \notin J} \varphi_j(s)} \qquad i \in J,$$

$$f_i(s) = \frac{s_i + \varphi_i(s)}{1 + \sum\limits_{j \notin J} \varphi_j(s)} \qquad i \notin J.$$

This implies, for every $s \in U$, that

$$f_i(s) \leq s_i \qquad \text{for all} \quad i \in J. \tag{A.17}$$

Let us assume that $f(U) \subseteq U$.

If we now define, starting with some $s^0 \in U$, a sequence $\left(s^k\right)_{k \in \mathbb{N}_0}$ in U via $s^{k+1} = f\left(s^k\right)$, then it follows from (A.17) that, for every $i \in J$, the sequence $\left(s_i^k\right)_{k \in \mathbb{N}_0}$ converges to some $s_i^* \in [0, 1]$.

We assume that also for every $i \notin J$ the sequence $\left(s_i^k\right)_{k \in \mathbb{N}_0}$ converges to some $s_i^* \in [0, 1]$, so that the sequence $\left(s^k\right)_{k \in \mathbb{N}_0}$ converges to some $s^* \in \overline{U} \subseteq S$ which is a fixed point of f and hence a Nash equilibrium.

In the special case where $J = \{1, \ldots, n\} \setminus \{i_0\}$ for some $i_0 \in \{1, \ldots, n\}$ this implies in particular that because of

$$f_{i_0}(s) = 1 - \sum_{i \neq i_0} f_i(s) \geq 1 - \sum_{i \neq i_0} s_i = s_{i_0},$$

the sequence $\left(s_{i_0}^k\right)_{k \in \mathbb{N}_0}$ converges to some $s_{i_0}^* \in [0, 1]$ and that

$$\begin{aligned} e_{i_0}^T A\, s^{*T} - s^* A\, s^{*T} &= \sum_{j=1}^n a_{i_0 j}\, s_j^* - \sum_{i=1}^n s_i^* \sum_{j=1}^n a_{ij}\, s_j^* \\ &= \sum_{i=1}^n s_i^* \left(\sum_{j=1}^n a_{i_0 j}\, s_j^* - \sum_{j=1}^n a_{ij}\, s_j^* \right) \\ &= \sum_{\substack{i=1 \\ i \neq i_0}}^n s_i^* \sum_{j=1}^n \left(a_{i_0 j} - a_{ij}\right)\, s_j^* \leq 0. \end{aligned}$$

Assumption:

$$a_{i_0 j} - a_{ij} > 0 \qquad \text{for all} \quad i \neq i_0 \quad \text{and} \quad j = 1, \ldots, n. \tag{A.18}$$

Then it follows that

$$s_i^* = 0 \qquad \text{for all} \quad i \neq i_0, \qquad \text{hence} \quad s_{i_0}^* = 1 \quad \text{and therefore} \quad s^* = e_{i_0}.$$

Let us demonstrate this by an example:

Let $n = 3$ and

$$A = \begin{pmatrix} 9 & 6 & 3 \\ 8 & 5 & 2 \\ 7 & 4 & 1 \end{pmatrix}.$$

Then we calculate

$$\begin{aligned} e_1 A s^T &= 9 s_1 + 6 s_2 + 3 s_3, \\ e_2 A s^T &= 8 s_1 + 5 s_2 + 2 s_3, \\ e_3 A s^T &= 7 s_1 + 4 s_2 + s_3, \\ s A s^T &= 9 s_1 + 6 s_2 + 3 s_3 - s_2 - 2 s_3. \end{aligned}$$

From this we obtain

$$\begin{aligned} e_1 A s^T - s A s^T &= s_2 + 2 s_3 > 0 \quad &&\text{for all} \quad s \in S \quad \text{with} \quad s_2 + 2 s_3 > 0 \\ & &&\iff s_1 - s_3 < 1, \\ e_2 A s^T - s A s^T &= -1 + s_2 + 2 s_3 < 0 \quad &&\text{for all} \quad s \in S \quad \text{with} \quad s_2 + 2 s_3 < 1 \\ & &&\iff s_3 < s_1, \\ e_3 A s^T - s A s^T &= -2 + s_2 + 2 s_3 < 0 \quad &&\text{for all} \quad s \in S \quad \text{with} \quad s_2 + 2 s_3 < 2 \\ & &&\iff 0 \leq 2 s_1 + s_2. \end{aligned}$$

This implies

$$\begin{aligned} \varphi_1(s) > 0 \qquad &\text{for all} \quad s \in U = \{ s \in S \mid 0 < s_3 < s_1 < 1 \}, \\ \varphi_2(s) = \varphi_3(s) = 0 \qquad &\text{for all} \quad s \in U \end{aligned}$$

and in turn, for every $s \in U$,

$$f_i(s) < s_i \qquad \text{for all} \quad i = 2, 3 \quad \text{and} \quad f_1(s) > s_1$$

which shows that (A.17) is satisfied for $i_0 = 1$.

Further, it follows that $f(U) \subseteq U$. The assumption (A.18) is also fulfilled.

In general, there exists a subsequence $(k_l)_{l \in \mathbb{N}_0}$ of $\mathbb{N}_0$ such that every sequence $\left(s_i^{k_l} \right)_{l \in \mathbb{N}_0}$ converges to some $s_i^* \in [0, 1]$ for every $i \notin J$. This implies

$$s_i^* = \frac{s_i^*}{1 + \sum\limits_{j \notin J} \varphi_j(s^*)} \qquad \text{for all} \quad i \in J.$$

If there exists some $i_0 \in J$ with $s_{i_0}^* > 0$, then it follows that

$$\varphi_j(s^*) = 0 \qquad \text{for all} \quad j \notin J \quad \text{and hence for all} \quad j \in \{1, \ldots, n\}$$

which implies that s^* is a Nash equilibrium.

Definition. *A Nash equilibrium $s^* \in S$ is called evolutionarily stable if $s\,A\,s^{*T} = s^*\,A\,s^{*T}$ for some $s \in S$ with $s \neq s^*$ implies that $s\,A\,s^T < s^*\,A\,s^T$.*

In words this means: If the average pay-off of s^* to s^* is the same as that of s^* for some $s \in S$ with $s \neq s^*$, then s cannot be a Nash equilibrium.

Assertion: If, for some $i_0 \in \{1, \ldots, n\}$, the condition (A.18) is satisfied, then e_{i_0} is an evolutionarily stable Nash equilibrium.
Condition (A.18) implies that

$$a_{i_0 i_0} > a_{i i_0} \qquad \text{for all} \quad i \neq i_0.$$

From this we infer, for every $s \in S$ with $s \neq e_{i_0}$, that

$$s\,A\,e_{i_0}^T = \sum_{i=1}^{n} s_i\,a_{i i_0} < a_{i_0 i_0} = e_{i_0}\,A\,e_{i_0}^T$$

which implies that e_{i_0} is an evolutionarily stable Nash equlibrium.

In [22] we have shown that for every evolutionarily stable Nash equilibrium $s^* \in S$ there exists some $\varepsilon^* > 0$ such that

$$s\,A\,s^T < s^*\,A\,s^T \qquad \text{for all} \quad s \in S \quad \text{with} \quad s \neq s^* \quad \text{and} \quad \|s - s^*\|_2 < \varepsilon^*.$$

Now let e_{i_0} be an evolutionarily stable Nash equilibrium. Then there is a relatively open set $U \subseteq S$ with $e_{i_0} \in U$ and

$$s\,A\,s^T < e_{i_0}\,A\,s^T \qquad \text{for all} \quad s \in U \quad \text{with} \quad s \neq e_{i_0}.$$

This implies

$$\varphi_{i_0}(s) > 0 \qquad \text{for all} \quad s \in U \quad \text{with} \quad s \neq e_{j_0}.$$

Assumption:

$$\varphi_i(s) = 0 \qquad \text{for all} \quad s \in U \quad \text{and} \quad i \neq i_0.$$

Then it follows that

$$\begin{aligned} f_i(s) &< s_i && \text{for all} \quad s \in U \quad \text{and} \quad i \neq i_0, \\ f_{i_0}(s) &> s_{i_0} && \text{for all} \quad s \in U \quad \text{and} \quad s \neq e_{i_0}. \end{aligned}$$

Now let us define a Lyapunov function $V : S \to \mathbb{R}$ by

$$V(s) = 1 - s_{i_0}, \qquad s \in S.$$

Then

$$V(e_{i_0}) = 0 \quad \text{and} \quad V(s) > 0 \qquad \text{for all} \quad s \in S \quad \text{with} \quad s \neq e_{i_0},$$
$$V(f(s)) - V(s) = 1 - f_{i_0}(s) - 1 + s_{i_0} < 0 \qquad \text{for all} \quad s \in U, \; s \neq e_{i_0}.$$

By Theorem 1.3 it therefore follows that e_{i_0} is an asymptotically stable fixed point of f.

In [22] we also consider the following dynamical treatment of the game. We assume that

$$s\,A\,s^T > 0 \qquad \text{for all} \quad s \in S$$

and define a mapping $f : S \to S$ by putting

$$f_i(s) = \frac{e_i\,A\,s^T}{s\,A\,s^T}\,s_i \qquad i = 1, \ldots, n \;\; \text{and} \;\; s \in S.$$

Obviously $s^* \in S$ is a fixed point of f, i.e., $f(s^*) = s^*$, if and only if

$$e_i\,A\,s^{*T} = s^*\,A\,s^{*T} \qquad \text{for all} \quad i \in S(s^*)$$

where

$$S(s^*) = \{i \in \{1, \ldots, n\} \mid s_i^* > 0\}.$$

From Lemma 1.1 in Section 1.2.5 it therefore follows that s^* is a fixed point of f, if s^* is a Nash equilibrium.

Now the question arises under which conditions this Nash equilibrium is an asymptotically stable fixed point of f. In order to answer this question we make use of Theorem 1.3. For this purpose we put

$$X(s^*) = \{s \in S \mid S(s) \supseteq S(s^*)\}$$

and define a function

$$V(s) = \sum_{i \in S(s^*)} s_i^* \left(\ln s_i^* - \ln s_i\right), \qquad s \in X(s^*).$$

Then it follows that

$$V(s) \geq 0 \quad \text{for all} \quad s \in X(s^*) \qquad \text{and} \qquad V(s) = 0 \;\Leftrightarrow\; s = s^*$$

(see Section 1.2.5).

Now let us assume that s^* is evolutionarily stable. Then, by Lemma 1.15 in Section 1.2.5, there exists some $\varepsilon > 0$ such that

$$s\,A\,s^T < s^*\,A\,s^T \qquad \text{for all} \quad s \in S \quad \text{with} \;\; ||s - s^*||_2 < \varepsilon \;\; \text{and} \;\; s \neq s^*.$$

Let us put

$$\mathcal{U}(s^*) = \{s \in X(s^*) \mid ||s - s^*|| < \varepsilon\}.$$

Further we put

$$Y(s^*) = \{ s \in X(s^*) \mid e_i A s^T = e_j A s^T \text{ for all } i, j \in S(s^*) \}.$$

Then it follows that

$$s A s^T < e_i A s^T \qquad \text{for all } \; i \in S(s^*) \text{ and all } s \in Y(s^*) \cap \mathcal{U}(s^*) \text{ and } s \neq s^*.$$

Further we obtain

$$\begin{aligned} V(f(s)) &= \sum_{i \in S(s^*)} s_i^* \left(\ln s_i^* - \ln e_i A s^T + \ln s A s^T - \ln s_i \right) \\ &= V(s) + \sum_{i \in S(s^*)} s_i^* \ln \left(\frac{s A s^T}{e_i A s^T} \right), \qquad s \in Y(s^*) \cap \mathcal{U}(s^*), \end{aligned}$$

which implies

$$V(f(s)) < V(s) \qquad \text{for all } \; s \in Y(s^*) \cap \mathcal{U}(s^*) \; \text{ with } \; s \neq s^*.$$

From Theorem 1.3 with $X = S$ and $G = Y(s^*) \cap \mathcal{U}(s^*)$ it then follows that s^* is an asymptotically stable fixed point of f.

B

Optimal Control in Chemotherapy of Cancer

B.1 The Mathematical Model and Two Control Problems

We describe the time dependent size of the tumor by a real valued function $T = T(t)$, $t \in \mathrm{I\!R}$ (t denotes the time), which we assume to be differentiable. The temporal development of this size (without treatment) we assume to be governed by the differential equation

$$\dot{T}(t) = f(T(t))\,T(t), \qquad t \in \mathrm{I\!R},$$

where the function $f : \mathrm{I\!R}_+ \to \mathrm{I\!R}_+$ is the growth rate of the tumor which is assumed to be in $\mathcal{C}^1(\mathrm{I\!R}_+)$ and to satisfy the condition

$$f'(T) < 0 \qquad \text{for all} \quad T \geq 0.$$

Let $[0, t_f]$ for some $t_f > 0$ be the time interval of treatment. The controlled growth of the tumor within this time interval is assumed to be described by the differential equation

$$\dot{T}(t) = \Big(f(T(t)) - \varphi(t)\,L(M(t))\Big)\,T(t), \qquad t \in (0, t_f)\,, \tag{B.1}$$

with the initial condition

$$T(0) = T_0 > 0 \tag{B.2}$$

at the beginning of the treatment.

The function $L = L(M)$, $M \geq 0$, denotes the destruction rate of the drug amount M and is assumed to satisfy the conditions

$$L \in \mathcal{C}^1(\mathrm{I\!R}_+)\,, \; L(0) = 0 \;\text{ and }\; L'(M) > 0 \qquad \text{for all} \quad M \geq 0.$$

S. Pickl and W. Krabs, *Dynamical Systems: Stability, Controllability and Chaotic Behavior*,
DOI 10.1007/978-3-642-13722-8_5, © Springer-Verlag Berlin Heidelberg 2010

The function $\varphi = \varphi(t),\ t \in [0, t_f]$ is a so called "resistance factor" that takes care of a possible resistance of the cancer cells against the drugs that is developed in the course of the treatment. It is assumed to be in $\mathcal{C}^1 [0, t_f]$ and to satisfy the conditions

$$\varphi(0) = 1,\ \dot{\varphi}(t) \leq 0 \qquad \text{for all} \quad t \in [0, t_f] \ \text{ and } \ \varphi(t_f) > 0.$$

The function $M = M(t),\ t \in [0, t_f]$, denotes the drug amount at the time t and is assumed to satisfy the differential equation

$$\dot{M}(t) = -\delta M(t) + V(t), \qquad t \in [0, t_f], \tag{B.3}$$

and the initial condition

$$M(0) = 0 \tag{B.4}$$

where $V(t)$ denotes the dosis that is administered per time unit at the time t and $\delta \geq 0$ is the decay rate of the drug. We assume the function $V = V(t)$ to be measurable and bounded and to satisfy the condition

$$0 \leq V(t) \leq A \qquad \text{for almost all} \quad t \in [0, t_f] \tag{B.5}$$

where $A > 0$ is a given constant.

In the sequel we consider the following two control problems:

Problem 1: Determine a control function $V \in L^\infty [0, t_f]$ such that under the conditions (B.1)–(B.5) and the end condition

$$T(t_f) = t_f \tag{B.6}$$

for some $T_f \in (0, T_0)$ the total amount of drugs administered which is given by

$$\int_0^{t_f} V(t)\ \mathrm{d}t$$

is minimized .

Problem 2: Determine a control function $V \in L^\infty [0, t_f]$ such that under the conditions (B.1)–(B.5) and

$$\int_0^{t} V(t)\ \mathrm{d}t = B \tag{B.7}$$

for some constant $B > 0$ the end value $T(t_f)$ is minimized.

If we define

$$y = \ln T \qquad \text{for} \quad T > 0,$$

then the differential equation (B.1) is transformed into

$$\dot{y}(t) = f\left(\mathrm{e}^{y(t)}\right) - \varphi(t)\, L(M(t)), \qquad t \in (0, t_f)\,, \tag{B.1'}$$

the initial condition (B.2) reads

$$y(0) = y_0 = \ln T_0 \tag{B.2'}$$

and the end condition (B.6) is given by

$$y\,(t_f) = y_f = \ln T_f\,. \tag{B.6'}$$

B.2 Solution of the First Control Problem

Let us repeat the problem to be solved: Determine a control function $V \in L^\infty[0, t_f]$ such that under the conditions (B.1'),(B.2'),(B.3)–(B.5),(B.6') the functional

$$F(y, M, V) = \int_0^{t_f} V(t)\,\mathrm{d}t$$

is minimized.

The Hamiltonian of this problem reads

$$H(t, y, M, V, p_1, p_2, \lambda_0) = (f(\mathrm{e}^y) - \varphi L(M))\, p_1 + (-\delta M + V) p_2 - \lambda_0 V.$$

Let $\left(\widehat{y}, \widehat{M}, \widehat{V}\right)$ be an optimal tripel; then, by the maximum principle (see A. D. Joffe and V. M. Tichomirov: *Theorie der Extremalaufgaben*, VEB Deutscher Verlag der Wissenschaften, Berlin 1979) there exists a number $\lambda_0 \geq 0$ and two functions $p_1, p_2 \in \mathcal{C}^1[0, t_f]$ with $(p_1, p_2, \lambda_0) \neq (0, 0, 0)$ and

$$\dot{p}_1(t) = -f'\left(\mathrm{e}^{\widehat{y}(t)}\right) \mathrm{e}^{\widehat{y}(t)} p_1(t), \qquad t \in [0, t_f],$$

$$\dot{p}_2(t) = \varphi(t)\, L'\left(\widehat{M}(t)\right) p_1(t) + \delta\, p_2(t), \quad t \in [0, t_f], \qquad p_2(t_f) = 0;$$

further we have

$$(p_2(t) - \lambda_0)\, \widehat{V}(t) = \max_{0 \leq V \leq A} (p_2(t) - \lambda_0)\, V \qquad \text{for almost all} \quad t \in [0, t_f].$$

If we put

$$g(t) = -f'\left(\mathrm{e}^{\widehat{y}(t)}\right) \mathrm{e}^{\widehat{y}(t)}, \qquad t \in [0, t_f],$$

then we obtain

$$\dot{p}_1(t) = g(t)\, p_1(t), \qquad t \in [0, t_f],$$

and therefore

$$p_1(t) = p_1(0)\, \mathrm{e}^{\int_0^t g(s)\,\mathrm{d}s}, \qquad t \in [0, t_f].$$

If we put

$$h(t) = \varphi(t)\, L'\left(\widehat{M}(t)\right) p_1(t), \qquad t \in [0, t_f],$$

then it follows that

$$\dot{p}_2(t) = \delta\, p_2(t) + h(t), \qquad t \in [0, t_f],$$

which implies

$$p_2(t) = \mathrm{e}^{\delta t}\left(p_2(0) + \int_0^t h(s)\, \mathrm{e}^{-\delta s}\,\mathrm{d}s\right), \qquad t \in [0, t_f].$$

From $p_2(t_f) = 0$ we infer

$$p_2(0) = -\int_0^{t_f} h(s)\,\mathrm{e}^{-\delta s}\,\mathrm{d}s$$

and therefore

$$p_2(t) = -\mathrm{e}^{\delta t}\int_t^{t_f} h(s)\,\mathrm{e}^{-\delta s}\,\mathrm{d}s, \qquad t \in [0, t_f]\,.$$

Now we distinguish three cases:

a) $p_1(0) = 0$; then $p_1(t) = 0$ for all $t \in [0, t_f]$ and therefore $h(t) = 0$ for all $t \in [0, t_f]$ which implies $p_2(t) = 0$ for all $t \in [0, t_f]$ and $\lambda_0 > 0$. Hence it follows that

$$p_2(t) - \lambda_0 < 0 \qquad \text{for all} \quad t \in [0, t_f]$$

and therefore

$$\widehat{V}(t) = 0 \qquad \text{for almost all} \quad t \in [0, t_f]\,.$$

This implies $\widehat{M}(t) = 0$ for all $t \in [0, t_f]$ and in turn

$$L\left(\widehat{M}(t)\right) = 0 \qquad \text{for all} \quad t \in [0, t_f]\,.$$

From this it follows that $\widehat{y}(t_f) > y_0 > y_{t_f}$ contradicting the requirement $y(t_f) = y_f$.

b) $p_1(0) > 0$; then it follows that $h(t) > 0$ for all $t \in [0, t_f]$ which implies $p_2(t) < 0$ for all $t \in [0, t_f)$ and therefore

$$p_2(t) - \lambda_0 < 0 \qquad \text{for all} \quad t \in [0, t_f)$$

which again leads to

$$\widehat{V}(t) = 0 \qquad \text{for almost all} \quad t \in [0, t_f]$$

being impossible as seen in a). Therefore we necessarily have

c) $p_1(0) < 0$; then it follows that $h(t) < 0$ for all $t \in [0, t_f]$ which implies $p_2(t) > 0$ for all $t \in [0, t_f)$.
α) Let $\lambda_0 = 0$; then it follows $p_2(t) - \lambda_0 > 0$ for all $t \in [0, t_f)$ and therefore

$$\widehat{V}(t) = A \qquad \text{for almost all} \quad t \in [0, t_f]\,. \tag{B.8}$$

β) Let $\lambda_0 > 0$; then it follows that $p_2(t_f) - \lambda_0 = -\lambda_0 < 0$.

Assumption:

$$\dot{p}_2(t) < 0 \qquad \text{for all} \quad t \in [0, t_f]\,. \tag{B.9}$$

This assumption implies that $p_2 = p_2(t)$, $t \in [0, t_f]$, is a strictly decreasing function with $p_2(0) > 0$ and $p_2(t_f) = 0$.

If $p_2(0) - \lambda_0 \leq 0$, then it follows that $p_2(t) - \lambda_0 < 0$ for all $t \in [0, t_f]$ which is impossible as shown in a). Therefore $p_2(0) - \lambda_0 > 0$; since $p_2(t_f) - \lambda_0 = -\lambda_0 < 0$ there exists exactly one $t_0 \in (0, t_f)$ such that

$$p_2(t) - \lambda_0 \begin{cases} > 0 & \text{for} \quad t \in [0, t_0)\,, \\ = 0 & \text{for} \quad t = t_0, \\ < 0 & \text{for} \quad t \in (t_0, t_f] \end{cases}$$

which implies

$$\widehat{V}(t) = \begin{cases} A & \text{for almost all} \quad t \in [0, t_0)\,, \\ 0 & \text{for almost all} \quad t \in (t_0, t_f]\,. \end{cases} \tag{B.10}$$

Result. *Under the assumption (B.9) an optimal control is necessarily of the form (B.8) or (B.10).*

Let us give a sufficient condition for (B.9) to hold true in the case $\lambda_0 > 0$. From the assumption on $\varphi = \varphi(t)$ it follows that

$$\varphi(t) > 0 \qquad \text{for all} \quad t \in [0, t_f]\,.$$

This then implies that

$$h(t) = \varphi(t)\, L'\left(\widehat{M}(t)\right) \widetilde{p}_1(t) < 0 \qquad \text{for all} \quad t \in [0, t_f]\,,$$

if $\lambda_0 > 0$. Because of

$$\dot{p}_2(t) = \delta\, p_2(t) + h(t), \qquad t \in [0, t_f]\,,$$

the assumption (B.9) is equivalent to

$$\delta\, p_2(t) + h(t) < 0 \qquad \text{for all} \quad t \in [0, t_f]$$

which is satisfied, if $\delta = 0$ where no decay of the drug takes place.

If $\delta > 0$, it follows from

$$p_2(t) = -\mathrm{e}^{\delta t} \int_t^{t_f} h(s)\, \mathrm{e}^{-\delta s}\, \mathrm{d}s, \qquad t \in [0, t_f]\,,$$

that

$$h(t) - \delta \int_t^{t_f} \mathrm{e}^{\delta(t-s)} h(s)\, \mathrm{d}s < 0, \qquad t \in [0, t_f]\,, \tag{B.11}$$

is necessary and sufficient for (B.9) to hold.

Since this condition is satisfied for $\delta = 0$, it is to be assumed that it is satisfied, if $\delta > 0$ is sufficiently small.

This can be quantified in the special case of the Gompertz growth law

$$f(T) = \lambda \ln \frac{\Theta}{T}, \qquad T > 0,$$

with constants $\lambda > 0$ and $\Theta > 0$ and a linear destruction rate

$$L(M) = c\,M, \qquad M \geq 0,$$

with a constant $c > 0$. In this case we obtain

$$g(t) = -f'\left(\mathrm{e}^{\widehat{y}(t)}\right)\mathrm{e}^{\widehat{y}(t)} = \lambda, \qquad t \in [0, t_f],$$

and

$$h(t) = c\,\varphi(t)\,\widetilde{p}_1(0)\,\mathrm{e}^{\lambda t}, \qquad t \in [0, t_f].$$

Condition (B.11) turns out to be equivalent to

$$\varphi(t)\,\mathrm{e}^{(\lambda-\delta)t} - \delta \int_t^{t_f} \varphi(s)\,\mathrm{e}^{(\lambda-\delta)s}\,\mathrm{d}s > 0 \qquad \text{for all} \quad t \in [0, t_f]. \tag{B.12}$$

If we put

$$\varphi(t) = \mathrm{e}^{-(\lambda-\delta)t}, \qquad t \in [0, t_f],$$

and assume $\lambda > \delta$, then it follows that $\varphi \in \mathcal{C}^1[0, t_f]$,

$$\varphi(0) = 1,\ \dot{\varphi}(t) < 0 \ \text{ and } \ \varphi(t) > 0 \qquad \text{for all} \quad t \in [0, t_f].$$

Further (B.12) turns out to be equivalent to

$$\left(1 - \delta\,(t_f - t) > 0 \ \text{ for all } \ t \in [0, t_f]\right) \quad \Longleftrightarrow \quad \delta\,t_f < 1.$$

B.3 Solution of the Second Control Problem

If we define

$$Q(t) = \int_0^t V(s)\,\mathrm{d}s, \qquad t \in [0, t_f],$$

then condition (B.7) turns out to be equivalent to

$$\dot{Q}(t) = V(t) \quad \text{for almost all} \quad t \in [0, t_f], \qquad Q(0) = 0,\; Q(t_f) = B. \quad \text{(B.13)}$$

Problem 2 then can be formulated in the following form: Find a function $V \in L^\infty[0, t_f]$ such that under the conditions (B.1'),(B.2'),(B.3)–(B.5) and (B.13) the functional

$$F(y, M, Q, V) = y(t_f) - y_0 = \int_0^{t_f} \left(f\left(\mathrm{e}^{y(t)}\right) - \varphi(t)\, L(M(t)) \right) \mathrm{d}t$$

is minimized.

The Hamiltonian of this control problem is given by

$$H(t, y, M, Q, V, p_1, p_2, p_3, \lambda_0) = (f(\mathrm{e}^y) - \varphi\, L(M))(p_1 - \lambda_0) + (V - \delta\, M) p_2 + V\, p_3.$$

Now let $\left(\widehat{y}, \widehat{M}, \widehat{Q}, \widehat{V}\right)$ be an optimal quadruple. Then by the maximum principle there exists a number $\lambda_0 \geq 0$ and three absolutely continuous functions p_1, p_2, p_3 on $[0, t_f]$ with $(\lambda_0, p_1, p_2, p_3) \neq (0,0,0,0)$ such that

$$\dot{p}_1(t) = -f'\left(\mathrm{e}^{\widehat{y}(t)}\right) \mathrm{e}^{\widehat{y}(t)} (p_1(t) - \lambda_0), \; t \in [0, t_f], \qquad p_1(t_f) = 0,$$

$$\dot{p}_2(t) = \varphi(t)\, L'\left(\widehat{M}(t)\right) (p_1(t) - \lambda_0) + \delta\, p_2(t), \; t \in [0, t_f], \qquad p_2(t_f) = 0,$$

$$\dot{p}_3(t) = 0 \quad \text{for almost all} \quad t \in [0, t_f]$$
$$\Longrightarrow p_3(t) = p_3 = \text{const.} \quad \text{for almost all} \quad t \in [0, t_f].$$

Further it is

$$(p_2(t) + p_3)\, \widehat{V}(t) = \max_{0 \leq V \leq A} (p_2(t) + p_3)\, V \qquad \text{for almost all} \quad t \in [0, t_f].$$

If we define $\widetilde{p}_1(t) = p_1(t) - \lambda_0$ and $g(t) = -f'\left(\mathrm{e}^{\widehat{y}(t)}\right) \mathrm{e}^{\widehat{y}(t)}$ for $t \in [0, t_f]$, then it follows that

$$\dot{\widetilde{p}}_1(t) = g(t)\, \widetilde{p}_1(t) \qquad \text{for all} \quad t \in [0, t_f]$$

and hence

$$\widetilde{p}_1(t) = \widetilde{p}_1(0)\, \mathrm{e}^{\int_0^t g(s)\,\mathrm{d}s}, \qquad t \in [0, t_f].$$

From $p_1(t_f) = 0$ we obtain

$$\widetilde{p}_1(t_f) = -\lambda_0 = \widetilde{p}_1(0)\, \mathrm{e}^{\int_0^{t_f} g(s)\,\mathrm{d}s}.$$

If we define

$$h(t) = \varphi(t)\, L'\left(\widehat{M}(t)\right)\, \widetilde{p}_1(t), \qquad t \in [0, t_f]\,,$$

then it follows that

$$\dot{p}_2(t) = \delta\, p_2(t) + h(t) \qquad \text{for} \quad t \in (0, t_f)\,.$$

This implies

$$p_2(t) = \mathrm{e}^{\delta t}\left(p_2(0) + \int_0^t \mathrm{e}^{-\delta s} h(s)\ \mathrm{d}s\right), \qquad t \in [0, t_f]\,.$$

From $p_2\,(t_f) = 0$ we obtain

$$p_2(0) = -\int_0^{t_f} \mathrm{e}^{-\delta s} h(s)\ \mathrm{d}s$$

which implies

$$p_2(t) = -\mathrm{e}^{\delta t} \int_t^{t_f} \mathrm{e}^{-\delta s} h(s)\ \mathrm{d}s, \qquad t \in [0, t_f]\,.$$

Now we distinguish two cases:

1) $\lambda_0 = 0$; then it follows that $p_1(t) = \widetilde{p}_1(t) = 0$ for all $t \in [0, t_f]$ which implies $h(t) = 0$ for all $t \in [0, t_f]$ and in turn

$$p_2(t) = 0 \qquad \text{for all} \quad t \in [0, t_f]\,.$$

Because of $(\lambda_0, p_1, p_2, p_3) \neq (0,0,0,0)$ it follows that $p_3 \neq 0$.

a) Let $p_3 < 0$; then it follows that

$$\widehat{V}(t) = 0 \qquad \text{for almost all} \quad t \in [0, t_f]$$

which is impossible because of (B.7) with $B > 0$.

b) Let $p_3 > 0$; then it follows that

$$\widehat{V}(t) = A \qquad \text{for almost all} \quad t \in [0, t_f] \tag{B.8}$$

which implies $At_f = B$.

2) $\lambda_0 > 0$; then it follows that $\widetilde{p}_1(t) < 0$ for all $t \in [0, t_f]$ which implies $h(t) < 0$ for all $t \in [0, t_f]$ and in turn $p_2(t) > 0$ for all $t \in [0, t_f)$ from which again we infer (B.8) and $At_f = B$, if $p_3 \geq 0$.

If $p_3 < 0$, we again make the assumption

$$\dot{p}_2(t) < 0 \qquad \text{for all} \quad t \in [0, t_f]\,. \tag{B.9}$$

Then $p_2 = p_2(t)$ is a strictly decreasing function from $p_2(0) > 0$ to $p_2\,(t_f) = 0$.

If $p_2(0) + p_3 \leq 0$, then $p_2(t) + p_3 < 0$ for all $t \in (0, t_f]$ implying

$$\widehat{V}(t) = 0 \qquad \text{for almost all} \quad t \in [0, t_f]$$

which is impossible (see above).

If $p_2(0) + p_3 > 0$, then there is exactly one $t_0 \in (0, t_f)$ with

$$\begin{aligned} &p_2(t) + p_3 > 0 \quad \text{for all} \quad t \in [0, t_0)\,, \qquad p_2\,(t_0) + p_3 = 0, \\ &p_2(t) + p_3 < 0 \quad \text{for all} \quad t \in (t_0, t_f]\,. \end{aligned}$$

As a consequence of this we obtain

$$\widehat{V}(t) = \begin{cases} A & \text{for almost all} \quad t \in [0, t_0)\,, \\ 0 & \text{for almost all} \quad t \in (t_0, t_f]\,. \end{cases} \tag{B.10}$$

where t_0 is the solution of the equation

$$\int_0^{t_0} A \ \mathrm{d}t = B \quad \Longleftrightarrow \quad A\,t_0 = B.$$

Result. *Under the assumption (B.9) an optimal control has necessarily the form (B.8) or (B.10). The form (B.8) is possible, if* $A t_f = B$. *The form (B.10) occurs, if* $A t_f > B$.

As in Section 2 one can show that the condition

$$h(t) = -\delta \int_t^{t_f} \mathrm{e}^{\delta(t-s)}\, h(s)\ \mathrm{d}s < 0, \qquad t \in [0, t_f]\,, \tag{B.11}$$

is necessary and sufficient for (B.9) to hold true in the case $\lambda_0 > 0$ and $p_3 < 0$.

Again in the special case of the Gompertz growth law for the untreated tumor and a linear destruction rate we can show for

$$\varphi(t) = \mathrm{e}^{-(\lambda-\delta)t}, \ t \in [0, t_f] \quad \text{with} \quad \lambda > \delta$$

that condition (B.11) is satisfied, if and only if $\delta\, t_f < 1$.

In addition the differential equation (B.1') reads

$$\dot{y}(t) = -\lambda\, y(t) + \lambda \ln \Theta - c\,\varphi(t)\, M(t), \qquad t \in (0, t_f)\,.$$

Under the initial condition (B.2') the unique solution is then given by

$$\begin{aligned} y(t) &= \mathrm{e}^{-\lambda t} \left(y_0 + \int_0^t \Big(\lambda \ln \Theta - c\,\varphi(s)\, M(s)\Big)\, \mathrm{e}^{\lambda s}\ \mathrm{d}s \right) \\ &= y_0\, \mathrm{e}^{-\lambda t} + \lambda \ln \Theta\, \left(1 - \mathrm{e}^{-\lambda t}\right) - c \int_0^t \mathrm{e}^{\lambda(s-t)}\, \varphi(s)\, M(s)\ \mathrm{d}s \\ &= y_0\, \mathrm{e}^{-\lambda t} + \lambda \ln \Theta\, \left(1 - \mathrm{e}^{-\lambda t}\right) - c\, \mathrm{e}^{-\lambda t} \int_0^t p(s)\, V(s)\ \mathrm{d}s, \qquad t \in [0, t_f]\,, \end{aligned}$$

where

$$p(s) = \mathrm{e}^{\delta s} \int_s^t \mathrm{e}^{(\lambda-\delta)\sigma} \varphi(\sigma)\, \mathrm{d}\sigma, \qquad s \in [0, t_f],$$

and the end value $y(t_f)$ reads

$$y(t_f) = y_0 \,\mathrm{e}^{-\lambda t_f} + \lambda \ln \Theta \left(1 - \mathrm{e}^{-\lambda t_f}\right) - c\,\mathrm{e}^{-\lambda t_f} \int_0^{t_t} p(t)\, V(t)\, \mathrm{d}t.$$

Problem 2 can now be equivalently formulated as follows:
Find a function $V \in L^\infty[0, t_f]$ with (B.5) and (B.7) such that

$$\int_0^{t_f} p(t)\, V(t)\, \mathrm{d}t \qquad \text{is maximized where} \quad p(t) = \mathrm{e}^{\delta t} \int_t^{t_f} \mathrm{e}^{(\lambda-\delta)s} \varphi(s)\, \mathrm{d}s.$$

Let $\widehat{V} \in L^\infty[0, t_f]$ be a solution of this problem (whose existence is ensured). then there exists some $\lambda_0 \in \mathbb{R}$ with

$$\int_0^{t_f} (p(t) - \lambda_0)\, \widehat{V}(t)\, \mathrm{d}t \geq \int_0^{t_f} (p(t) - \lambda_0)\, V(t)\, \mathrm{d}t$$
$$\text{for all} \quad V \in L^\infty[0, t_f] \quad \text{with} \quad (B.5).$$

This implies

$$\widehat{V}(t) = \begin{cases} A, & \text{if} \quad p(t) - \lambda_0 > 0, \\ 0, & \text{if} \quad p(t) - \lambda_0 < 0. \end{cases}$$

If $\lambda_0 \leq 0$, it follows that

$$p(t) - \lambda_0 > 0 \qquad \text{for all} \quad t \in [0, t_f)$$

which implies

$$\widehat{V}(t) = A \quad \text{for all} \quad t \in [0, t_f) \qquad \text{and} \qquad A t_f = B.$$

If $\lambda_0 > 0$, we again put

$$\varphi(t) = \mathrm{e}^{-(\lambda-\delta)t}, \ t \in [0, t_f], \qquad \text{with} \quad \lambda > \delta.$$

Then it follows that

$$p(t) = (t_f - t)\, \mathrm{e}^{\delta t}, \qquad t \in [0, t_f],$$

hence

$$\dot{p}(t) = (\delta\, (t_f - t) - 1)\, \mathrm{e}^{\delta t} < 0 \qquad \text{for all} \quad t \in [0, t_f],$$

if and only if $\delta\, t_f < 1$.

Under this condition $p = p(t)$ is a strictly decreasing function with $p(0) = t_f$ and $p(t_f) = 0$.

If $t_f \leq \lambda_0$, then $p(t) - \lambda_0 < 0$ for all $t \in (0, t_f]$ which implies

$$\widehat{V}(t) = 0 \qquad \text{for all} \quad t \in (0, t_f]$$

which is impossible.

If $t_f > \lambda_0$, then $p(0) - \lambda_0 > 0$ and $p(t_f) - \lambda_0 = -\lambda_0 < 0$ and there exists exactly one $t_0 \in (0, t_f)$ such that

$$\begin{aligned} p(t) - \lambda_0 > 0 \quad &\text{for all} \quad t \in [0, t_0), \qquad p(t_0) - \lambda_0 = 0, \\ p(t) - \lambda_0 < 0 \quad &\text{for all} \quad t \in (t_0, t_f] \end{aligned}$$

which implies

$$\widehat{V}(t) = \begin{cases} A, & \text{if} \quad t \in [0, t_0), \\ 0, & \text{if} \quad t \in (t_0, t_f], \end{cases}$$

and t_0 is the solution of the equation

$$\int_0^{t_0} A \, \mathrm{d}t = B \quad \Longleftrightarrow \quad A t_0 = B.$$

In addition we obtain

$$\int_0^{t_f} p(t) \, \widehat{V}(t) \, \mathrm{d}t = \int_0^{t_0} p(t) \, A \, \mathrm{d}t = A \int_0^{t_0} (t_f - t) \, \mathrm{e}^{\delta t} \, \mathrm{d}t. \tag{B.12}$$

Summarizing every solution of the above problem is the form

$$\widehat{V}(t) = \begin{cases} A, & \text{if} \quad t \in [0, t_0], \\ 0, & \text{if} \quad t \in (t_0, t_f], \quad \text{for some} \quad t_0 \in (0, t_f] \end{cases}$$

and the extreme value is given by (B.12).

Finally we consider the problem of determining a function $V \in L^\infty [0, t_f]$ which satisfies (B.5) and

$$\int_0^{t_f} p(t) \, V(t) \, \mathrm{d}t = A \int_0^{t_0} (t_f - t) \, \mathrm{e}^{\delta t} \, \mathrm{d}t$$

and which minimizes $\displaystyle\int_0^{t_f} V(t) \, \mathrm{d}t$.

Let $V^* \in L^\infty [0, t_f]$ be a solution of this problem. Then it is necessarily of the form (see Section 2)

$$V^*(t) = \begin{cases} A, & \text{if} \quad t \in [0, t_0^*], \\ 0, & \text{if} \quad t \in (t_0^*, t_f], \quad \text{for some} \quad t_0^* \in (0, t_f]. \end{cases}$$

This implies

$$\int_0^{t_f} p(t) \, V^*(t) \, \mathrm{d}t = A \int_0^{t_0^*} (t_f - t) \, \mathrm{e}^{\delta t} \, \mathrm{d}t = A \int_0^{t_0} (t_f - t) \, \mathrm{e}^{\delta t} \, \mathrm{d}t,$$

hence $t_0^* = t_0$ and $V^*(t) = \widehat{V}(t)$ for all $t \in [0, t_f]$.

B.4 Pontryagin's Maximum Principle

The two control problems considered in Sections B.1–B.3 are special cases of the following situation: Given a system of differential equations of the form

$$\dot{x}_i(t) = f_i(x(t), u(t)), \qquad t \in (t_0, t_1), \quad \text{for} \quad i = 1, \dots, n \tag{B.13}$$

with initial conditions

$$x_i\,(t_0) = x_{0i}, \qquad i = 1, \dots, n, \tag{B.14}$$

and conditions

$$x_{i_k}\,(t_1) = x_{1i_k}, \qquad k = 1, \dots, r(\leq n), \tag{B.15}$$

where $x(t) = (x_1(t), \dots, x_n(t))$, $u(t) = (u_1(t), \dots, u_m(t))$, $t \in [t_0, t_1]$, $f_i : \mathbb{R}^n \times \mathbb{R}^m \to \mathbb{R}$, $i = 1, \dots, m$, with $f_i \in \mathcal{C}\,(\mathbb{R}^n \times \mathbb{R}^m)$ and $f_i(\cdot, u) \in \mathcal{C}^1\,(\mathbb{R}^n)$ for every $u \in \mathbb{R}^m$. The functions $u_i = u_i(t)$ for $t \in [t_0, t_1]$ and $i = 1, \dots, m$ are considered as control functions and are assumed to be in the space $L^\infty\,[t_0, t_1]$. Further we assume that

$$u(t) \in \mathcal{U} \subseteq \mathbb{R}^m \qquad \text{for almost all} \quad t \in [t_0, t_1] \tag{B.16}$$

where $\mathcal{U}$ is a non-empty subset of $\mathbb{R}^m$.

Finally we consider a functional

$$F(x(\cdot), u(\cdot)) = \int_{t_0}^{t_1} f_0(x(t), u(t))\; dt. \tag{B.17}$$

The control problem now consists of determining some $u \in L^\infty\,([t_0, t_1]\,, \mathbb{R}^m)$ which satisfies (B.16) such that the functional (B.17) is minimized where $x = x(t)$, $t \in [t_0, t_1]$, is the corresponding solution of (B.13),(B.14),(B.15) in the space of absolutely continuous $n-$vector functions on $[t_0, t_1]$. Let us define a Hamilton function (or Hamiltonian) by

$$H\,(t, x, u, p, \lambda_0) = \sum_{j=1}^{n} p_j f_j(x, u) - \lambda_0 f_0(x, u)$$

where $p \in \mathbb{R}^n$, $\lambda_0 \in \mathbb{R}_+$. The we have the following

Maximum Principle: Let $(\widehat{x}, \widehat{u})$ be a solution of the above control problem. Then there exists a number $\lambda_0 \geq 0$ and n absolutely continuous functions $p_1 = p_1(t), \ldots, p_n = p_n(t)$, $t \in [t_0, t_1]$, with $(p_1, \ldots, p_n, \lambda_0) \neq (0, \ldots, 0) \in \mathbb{R}^{n+1}$ such that

$$\dot{p}_i(t) = -H_{x_i}\left(t, \widehat{x}(t), \widehat{u}(t), p(t), \lambda_0\right) \qquad (p(t) = (p_1(t), \ldots, p_n(t)))$$

$$= -\sum_{j=1}^{n} p_j f_{jx_i}\left(\widehat{x}(t), \widehat{u}(t)\right) + \lambda_0 f_{0x_i}\left(\widehat{x}(t), \widehat{u}(t)\right)$$

$$\text{for almost all} \quad t \in (t_0, t_1), \; i = 1, \ldots, n,$$

$$p_i(t_1) = 0 \qquad \text{for all} \quad i \notin \{i_1, \ldots, i_k\},$$

and

$$H\left(t, \widehat{x}(t), \widehat{u}(t), p(t), \lambda_0\right) = \max_{u \in \mathcal{U}} H\left(t, \widehat{x}(t), u, p(t), \lambda_0\right)$$

$$\text{for almost all} \quad t \in [t_0, t_1].$$

For the proof we refer to: A. D. Joffe and V. M. Tichomirov: *Theorie der Extremalaufgaben*, VEB Deutscher Verlag der Wissenschaften, Berlin 1979.

C

List of Authors

Prof. Dr. rer.nat. Werner Krabs

1934	born in Hamburg-Altona
1954–1959	study of mathematics, physics and astronomy at the University of Hamburg, diploma in mathematics
1963	PhD-thesis
1967/1968	visiting assistant professor at the University of Washington in Seattle
1968	habilitation in applied mathematics at the University of Hamburg
1970–1972	professor at the RWTH Aachen
1971	visiting associate professor at the Michigan State University in East Lansing
1972	professor at the TH Darmstadt
1977	visiting full professor at the Oregon State University in Corvallis
1979–1981	vice-president of the TH Darmstadt
1986–1987	chairman of the Society for Mathematics, Economy and Operations Research.

Prof. Dr. Stefan Pickl

Studied mathematics, electrical engineering, pedagogy and philosophy at TU Darmstadt and EPFL Lausanne 1987–93. Dipl.-Ing. 1993, Doctorate 1998 with award. Assistant Professor at Cologne University (Dr. habil. 2005; venia legendi "Mathematics"). Chair for Operations Research at UBw Munich since 2005.

Visiting Professor at University of New Mexico (USA), University Graz (Austria), Naval Postgraduate School Monterey, University of California at Berkeley. Visiting scientist at SANDIA, Los Alamos National Lab, Santa Fe Institute for Complex Systems and MIT. Associated with Centre for the Advanced Study of Algorithms (CASA, USA), with Center for Network Innovation and Experimentation (CENETIX, USA). Vice-chair of EURO working group Experimental Economics, chair of advisory board of the German Society for Operations Research (GOR), advisory board of the German Society for Disaster and Emergency Management. International Research Programs, 80 publications, international Best-Paper-Awards 2003, 2005, 2007. Foundation of COMTESSA (Competence Center for Operations Research, Management of Intelligent Engineered Secure Systems & Algorithms) and the Academy for Highly Gifted Pupils in Munich.

References

1. Agarwal, R. T.: *Difference Equations and Inequalities.* Marcel Dekker, Inc., New York, Basel, Hongkong (1992).
2. Banks, J., Brooks, J., Cairns, G., Davis, G., and Stacey, P.: *On Devaney's Definition of Chaos.* American Mathematical Monthly **99**, 332–334 (1992).
3. Birkhoff, G. D.: *Dynamical Systems.* American Math. Society Colloqu. Publ., Providence (1927).
4. Burger, E.: *Einführung in die Theorie der Spiele.* Walter de Gruyter & Co.,Berlin (1966).
5. Devaney, R. L.: *An Introduction to Chaotic Dynamical Systems.* Addison-Wesley (1989).
6. Gantmacher, F. R.: *Application of the Theory of Matrices.* Interscience Publishers, New York-London-Sydney (1959).
7. Hale, J. K.: *Ordinary Differential Equations.* Wiley-Interscience (1969).
8. Hale, J. K. (editor): *Studies in Ordinary Differential Equations.* **14**, published by the Mathematical Association of America (1977).
9. Hurwitz, A.: *Über die Bedingungen, unter welchen eine Gleichung nur Wurzeln mit negativen reellen Teilen besitzt.* Math. Ann. **46**, 273–284 (1895).
10. Janková, K., and Smital, J.:*A Characterization of Chaos.* Bull. Austral. Math. Soc. **34**, 283–292 (1986).
11. Kirchgraber, U.: *Mathematik im Chaos. Ein Zugang auf dem Niveau der Sekundarstufe II.* Mathematische Semesterberichte **39**, 43–68 (1992).
12. Knobloch, H. W., and Kappel, F.: *Gewöhnliche Differentialgleichungen.* Verlag B. G. Teubner, Stuttgart (1974).
13. Krabs, W.: *Optimal Control of Undamped Linear Vibrations.* Heldermann Verlag (1995).
14. Krabs, W.: *Dynamische Systeme: Steuerbarkeit und Chaotisches Verhalten.* Verlag B. G. Teubner, Stuttgart-Leipzig (1998).
15. Krabs, W.: *On Local Controllability of Time-Discrete Dynamical Systems into Steady States.* Journal of Difference Equations and Applications **8**, 1–11 (2002).
16. Krabs, W.: *Stability and Controllability in Non-Autonomous Time-Discrete Dynamical Systems.* Journal of Difference Equations and Applications **8**, 1107–1118 (2002).
17. Krabs, W.: *On Local Fixed Point Controllability of Nonlinear Difference Equations.* Journal of Difference Equations and Applications **9**, 827–832 (2003).

18. Krabs, W.: *A General Predator-Prey Model.* Mathematical and Computer Modelling of Dynamical Systems **9**, 387–401 (2003).
19. Krabs, W.: *A Dynamical Method for the Calculation of Nash-Equilibria in $n-$Person Games.* Journal of Difference Equations and Applications **11**, 919–932 (2005).
20. Krabs, W.: *Stability in Predator-Prey Models and Discretization of a Modified Volterra-Lotka-Model.* Mathematical and Computer Modelling of Dynamical Systems **12**, 577–588 (2006).
21. Krabs, W.: *On Local Reachability and Controllability of Nonlinear Difference Equations.* Mathematical and Computer Modelling of Dynamical Systems **12**, 487–494 (2006).
22. Krabs, W., and Pickl, St.: *Analysis, Controllability and Optimization of Time Discrete Systems and Dynamical Games.* Lecture Notes in Economics and Mathematical Systems **529**, Springer-Verlag, Berlin, Heidelberg (2003).
23. Krause, U., and Nesemann, T.: *Differenzengleichungen und diskrete dynamische Systeme.* B. G. Teubner, Stuttgart, Leipzig (1995).
24. Lakshmikantham, V., and Frigiante, D.: *Theory of Difference Equations: Numerical Methods and Applications.* Academic Press, Inc., Boston et al. (1988).
25. La Salle, J. P.: *The Stability and Control of Discrete Processes.* Springer Verlag, New York-Berlin-Heidelberg-London-Paris-Tokio (1986).
26. Li, J.: *Das dynamische Verhalten diskreter Evolutionsspiele.* Shaker Verlag, Aachen (1999).
27. Li, T. Y., and Yorke, J. A.: *Period Three Implies Chaos.* American Mathematical Monthly **85**, 985–992 (1975).
28. Mielke, A.: *Topological Methods for Discrete Dynamical Systems.* GAMM-Mitteilungen **2**, 19–37 (1990).
29. Misiurewicz, M.: *Horseshoes for Mappings of the Interval.* Bulletin de l'Académie Polonaise des Sciences **XXVII(2)**, 166–168 (1979).
30. Misiurewicz, M., and Szlenk, W.: *Entropy of Piecewise Monotonic Mappings.* Studia Matematica **LXVII**, 45–63 (1980).
31. Nash, J.: *Non-Cooperative Games.* Annals of Mathematics **54**, 286–295 (1951).
32. Pickl, St.: *Der $\tau-$value als Kontrollparameter. Modellierung und Analyse eines Joint-Implementation Programmes mithilfe der dynamischen kooperativen Spieltheorie und der diskreten Optimierung.* Shaker Verlag, Aachen (1999).
33. Sell, G. R.: *Topological Dynamics and Ordinary Differential Equations.* Van Nostrand Reinhold Company, London (1971).
34. Smital, J.: *Chaotic Functions with Zero Topological Entropy.* Transactions of the American Mathematical Society **297**, 269–282 (1986).

Index

$\Omega-$controllable, 76
$\mathbb{R}^m-$controllable, 77
$\mathbb{R}^m-$null-controllable, 83
$n-$person game, 191

admissible control functions, 120, 127, 129, 131
alpha limit set, 3
approximation problem, 87
asymptotical stability, 103
asymptotically stable, 12, 13, 19, 21, 22, 33, 34, 42, 44, 46, 48–54, 56, 64–69, 121–123, 138, 139, 207
asymptotically stable equilibrium state, 24, 25
asymptotically stable fixed point, 55, 56, 58, 207, 208
asymptotically stable rest point, 17, 19
attractive, 50, 51, 66, 67, 69
attractor, 41–45, 64, 65, 100, 138, 139, 175, 178, 204
autonomous, 35, 145
autonomous system, 13
autonomous time-discrete dynamical system, 60
average pay-off, 202

bi-matrix game, 31
Bi-Matrix Games, 196
Bouwer's fixed point theorem, 189
Brouwer's fixed point theorem, 192, 194, 203

Cantor set, 152, 178, 179
Cayley-Hamilton, 114
Chaos in the Sense of Devaney, 145
Chaos in the Sense of Li and Yorke, 168
chaos in the sense of Li and Yorke, 174, 187
chaotic behavior of dynamical systems, 186
chaotic behavior of one-dimensional continuous mappings, 186
chaotic in the sense of Devaney, 176, 178, 180, 182, 183, 186
chaotic in the sense of Li and Yorke, 174
chaotic mapping, 186
completely $\mathbb{R}^m-$controllable, 78
completely $\Omega-$controllable, 76
completely $\mathbb{R}^4-$controllable, 79
control function, 75, 83, 84, 99, 100, 106, 109, 120, 129, 131, 133, 134, 210, 211
Controllability of Linear Systems, 76
controllability of time-discrete dynamical systems, 185
controllable set, 101, 135
controlled system, 103, 106, 107, 123, 138
cost functional, 104
cyclic, 38

decay rate, 210
destruction rate, 209, 214, 217
difference equation, 128
difference equations, 99, 131
differential equation, 79, 84, 128, 209, 210

Direct Method of Lyapunov, 12
direct method of Lyapunov, 13, 184
Discretization, 55, 57
discretization, 54
disorder-chaos, 163, 174, 187
disorder-chaotic, 161, 162, 166, 167
Dual Problem, 91
dual problem, 91, 92, 96
dynamical system, 1, 3, 12, 17, 184
dynamical systems, 1

Emission Reduction Model, 58
emission reduction model, 107, 123
equilibrium, 6
equilibrium state, 20, 22, 23
evolution matrix game, 30
evolutionarily stable, 33, 34, 206, 207
evolutionarily stable Nash equilibrium, 32, 33

feedback controls, 138
fixed point, 38, 42, 44, 45, 49, 59, 99, 101, 103, 109, 121, 123, 129, 131, 134, 140, 189, 191, 192, 194, 197, 200, 203, 204, 207
Fixed Point Controllability, 99
fixed point controllability, 119
fixed point equation, 118
Fixed Point Theorem of Kakutani, 189
fixed point theorem of Kakutani, 189, 190
flow, 1, 6, 10–13

General Predator-Prey Model, 20
general predator-prey models, 184
global attractor, 110
globally asymptotically stable, 53
globally attractive, 86
Gompertz growth, 214, 217
gradient projection method, 117

halftrajectory, 3
Hamiltonian, 211, 215

identity property, 1, 2, 185
implicit function theorem, 102, 136, 137, 143
Interacting Logistic Growth, 57
invariant, 4, 6, 8, 36–40, 175
invariant subset, 4, 39, 44
invariantly connected, 38, 40
inverse function theorem, 102, 136, 137, 140

Kakutani's fixed point theorem, 192
Kalman condition, 77–79, 81, 82, 84–86, 110–112, 114, 116, 124

Lagrangean multiplier rule, 92
limit point, 12
limit set, 2, 35–38, 60, 175, 178
linear system, 109
Linearisation, 45
local controllability, 83
Local Fixed Point Controllability, 131
localization of limit sets, 184
locally controllable, 83
locally restricted controllable, 84, 85
Lyapunov, 1
Lyapunov function, 11, 13, 16, 17, 27, 29, 39, 40, 42, 44, 53, 62, 64, 65, 139, 184, 203, 206
Lyapunov's Method, 40
Lyapunov's method, 23

Marquardt's algorithm, 104
matrix game, 202
maximum principle, 211, 215
minimal, 6, 7
minimal subset, 6, 8
minimum norm control, 95, 96, 98
Minimum Norm Problem, 91
minimum norm problem, 91, 93, 94, 96
minimum time, 89, 95, 98
movement, 5, 10
moving linear pendulum, 78

Nash equilibrium, 31–33, 191–194, 196–200, 202, 204, 206, 207
non-autonomous linear systems, 184
non-autonomous systems, 184
non-autonomous time-discrete dynamical system, 60, 66, 99, 131, 137
non-linear planar pendulum, 128
nonlinear pendulum with movable suspension point, 84
Normality Condition, 90

normality condition, 91, 93, 95–98
null-controllability, 124

omega limit set, 3–5, 8, 12
optimal control, 213
Optimal Control in Chemotherapy of Cancer, 209
optimization problem, 104
orbit, 2, 11, 35

pay-off, 191
pay-off functions, 192, 196
pay-off matrix, 202
period, 38
period of the movement, 6
period point, 145–147, 166, 168, 172, 174
period points, 157
periodic, 6, 38
periodic movement, 7, 8
periodic orbit, 7, 8, 10, 12
plane pendulum, 8
Poincaré, 1
Poincaré-Bendixon theory, 184
positive definite, 13, 15, 17, 24, 41, 42, 44, 53, 63–65, 139
positive halftrajectory, 5
positive semi-orbit, 12
positively compact, 5, 10–12
positively compact movement, 5, 6
positively compact orbit, 8
positively invariant, 36, 38, 41, 44, 45
predator prey model, 103, 126
Predator-Prey Model, 10, 55
Problem of Controllability, 134
problem of fixed point controllability, 100
Problem of Null-Controllability, 109
problem of null-controllability, 110, 111, 116
Process of Hemo-Dialysis, 71

Reachability, 140
reachable, 140, 141
rest point, 6–8, 10–13, 15, 18–20, 33, 55, 76, 83, 86
Restricted Null-Controllability, 80
restricted null-controllability, 81, 87–89, 95, 98
restricted null-controllable, 82, 84

semigroup property, 1, 2, 185
sensitively depending on initial values, 146
set of period points, 145, 149, 150, 181
set of periodic points, 186
shift-mapping, 148, 157, 161, 178, 186, 187
Smale's horse-shoe-mapping, 176
Solenoid, 180
stability of fixed points, 184
Stabilization, 121, 137
stabilization, 138
stable, 12, 13, 40–42, 44, 50, 53, 63, 64, 66–69, 138, 139
stable equilibrium state, 22, 25, 27
stable fixed point, 52, 53
stable rest point, 16, 17
state function, 99, 131
strange attractor, 176, 180
strange attractors, 187
strategy sets, 191, 194, 196
subset, 4
subshift, 163
symmetric bi-matrix games, 200
system of differential equations, 75

Theorem of Sarkovski, 170, 174
time-continuous, 3
time-continuous dynamical system, 13
time-continuous dynamical systems, 184
time-discrete, 1
time-discrete autonomous dynamical system, 121
time-discrete dynamical system, 35, 54, 58, 156
time-discrete flow, 35
time-minimal control, 96, 98
time-minimal controllability, 185
time-minimal null-control, 95
time-minimal restricted null-controllability, 90
topological conjugacy, 187
topological entropy, 159, 161, 163, 165, 174, 187
topological transitivity, 186
topologically conjugated, 150, 156, 180, 183, 187

topologically transitive, 145, 147, 149, 150, 157, 181
trajectory, 2, 4

uncontrolled system, 75, 83, 86, 99–101, 103, 105, 109, 110, 123, 129, 131, 134, 140
unstable, 46, 48–50, 122, 123
unstable fixed point, 55, 57
upper semi-continuous, 189

Kolibri 360 GmbH
Ferdinand-Jühlke-Straße 7
99095 Erfurt, Deutschland
produktsicherheit@kolibri360.de